Barton S Lobdell

The Telluric Manual

A Guide to the Study of Swigert's Lunar Tellurian

Barton S Lobdell

The Telluric Manual
A Guide to the Study of Swigert's Lunar Tellurian

ISBN/EAN: 9783337371494

Printed in Europe, USA, Canada, Australia, Japan

Cover: Foto ©berggeist007 / pixelio.de

More available books at **www.hansebooks.com**

A GUIDE

TO THE

Study of Swigert's Lunar Tellurian

BY

B. S. LOBDELL, M. S.

Published by
CENTRAL SCHOOL SUPPLY HOUSE.
CHICAGO, ILL.

PREFACE.

This little book is designed to serve as a guide to the study of the Tellurian. The language used is simple, and the experiments with the Tellurian will enable the pupil to readily understand some of the phenomena regarding the sun, earth, moon, and the stars that have attracted his attention and become a mystery to him.

The experiments will tend to arouse the curiosity and stimulate the observation of the pupil, and will give a new meaning to all that he may learn in the study of Geography. Latitude will mean more to him than "distance north or south of the equator;" and, by learning the location of places, he will know, by induction, facts about them that would never occur to him without having seen the demonstrations that the Tellurian gives.

A careful study of places, as outlined in the first chapters of the Manual, is expected. After this the phenomena caused by the rotation of the earth on her axis, the inclination of the earth's axis, and the revolution of the earth around the sun are to be studied. This will teach the pupil the intimate relation that exists between latitude, the length of days and nights, and the distribution of light and heat. By easy steps he is led to fully understand all that is said in the chapter on Climate.

The study of the Tellurian as herein directed will be found to possess much value in acquiring a knowledge of Physical Geography.

Hitherto the globe has been presented as an abstract study. The author of the Telluric Manual invites your attention to a new treatment of this subject, and begs your indulgence.

CONTENTS.

CHAPTER		PAGE
I.	The Tellurian	5
II.	The Beginning	12
III.	Directions and Locations	17
IV.	Meridians and Longitude	26
V.	Parallels and Latitude	34
VI.	Study of the Globe's Surface	39
VII.	Longitude and Time	44
VIII.	Days and Nights	53
IX.	Yearly Revolution of the Earth	60
X.	Length of Days and Nights	66
XI.	(1) How to Compute the Length of Days and Nights (2) Twilight	75
XII.	Distribution of Light and Heat	85
XIII.	Climate	93
XIV.	The Moon	104
XV.	The Sun	124
XVI.	Eclipses	132
XVII.	The Calendar	145
	The Appendix	154

ILLUSTRATIONS.

	PAGE
Analemma	168
Compass, the points of	21
Crest and Trough of Tides	113
Day and Night	75
Day Hemisphere (in midsummer)	66
Eclipse, annular	141
Eclipse, partial solar	140
Eclipse, solar	138
Earth, a sectional view of	14
Figure, illustrating apparent rotation of the Heavens	156
Figure, showing orbit of moon and length of earth's shadow	134
Figures, showing orbits of sun and moon	132
Figures, showing relative size of earth and moon	133
Great Dipper, the, and pole star	20
Light and Heat, distribution of	86
Longitude and Latitude	39
Lunar Tellurian	7
Meridians	26
Nodes of the Moon	117
Oblique Rays of Sun on Slopes	89
Parallels	34
Polar Projection, showing east and west longitude	50
Tellurian, with day and night circles	10
Tides	111
Twilight	82
Zodiac, etc.	161

"The works of God are fair for naught,
 Unless our eyes, in seeing,
See hidden in the thing the thought
 That animates its being."

CHAPTER I.

THE TELLURIAN.

To get an accurate idea of the relative extent of bodies of land and water, we are obliged to study the surface of a *globe*, for in none of our school textbooks have we maps drawn to the same scale; and, when we say that one country is so many times the size of another, it has but little meaning compared with the truth as it appeals to the eye when the two countries are shown side by side.

The same is true of *latitude*. In comparing the latitude of New York and London, the pupil has in mind a map of the United States and a map of Europe. These cities are about the same distance from the top of the map, and to say that one is 41 degrees north latitude, and the other 52 degrees north latitude, does but little in removing the wrong impression that the maps have conveyed.

Placing the meridian circle upon New York, and slowly revolving the globe, the bodies of land and water, and the cities passing under the circle, will tell the story of latitude in a way that the youngest pupil can understand, and that the ablest teacher cannot make more easy.

Comparative Latitude—which means a comparison of the length of the days and nights, and of the seasons of the year—assists the pupil in classifying climate and productions, which in turn tells him much about the classes of people that he may expect to meet in foreign lands, and so on with the entire knowledge that we get by the study of geography.

It is the province of this book, studied in connection with the Tellurian, to make the work more a matter of experience, and to assist the pupil in observing conditions that directly affect every living thing upon the face of the earth. Much time will in this way be saved, and the study of geography be made more attractive.

Begin the study of the globe by locating the different bodies of land and water, continents and oceans, islands and seas and lakes, then rivers and towns.

Begin with the nearest city, and locate all of the principal cities, telling what nations own them.

Begin with New York, and locate all seaports. Tell how you would make a journey to each, etc.

We will now study the parts of the Tellurian that will be most helpful in gaining a concise knowledge of the earth's surface, her movements, and her relations to the moon and the sun, using the names given in the accompanying illustration

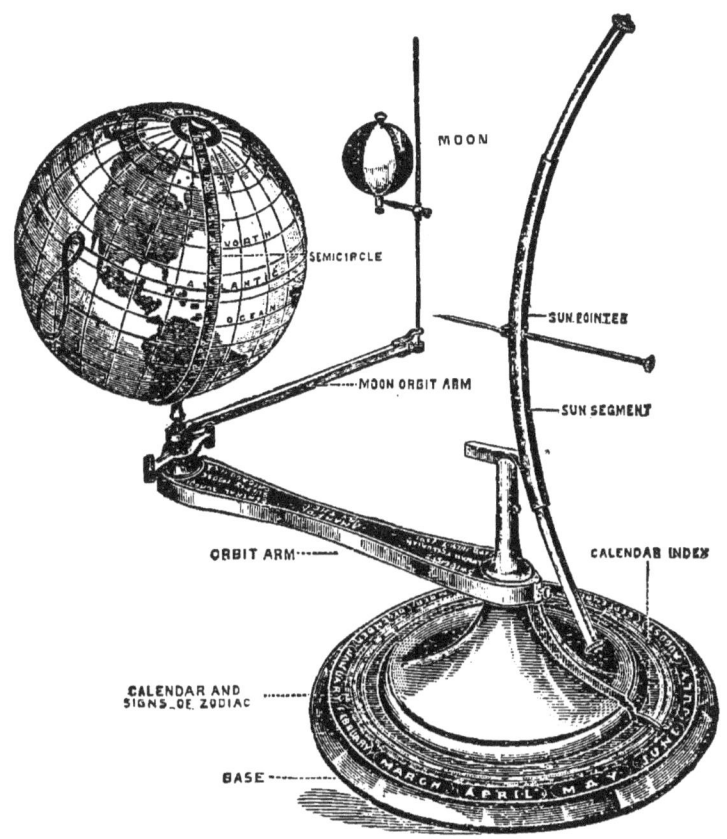

The *Orbit Arm* is used to revolve the earth about the sun in her yearly revolution.

The *Calendar Index* shows the positions of the earth in her orbit, by pointing at the names of the months and the days of the months arranged around the base of the Tellurian.

The *Base* of the Tellurian represents a belt in the Celestial Sphere, called the *Zodiac*. This belt is divided into twelve parts, called *Signs*, and their names will be found in Chapter IX. If you could stand at the center of the solar system and watch the earth, she would appear to be in one of these signs; and, if you could watch the earth for a year, she would appear to move from one of these signs to another until the entire circuit had been made. The *Index* shows the day and the month.

The larger globe, or ball, represents the *Earth*. We shall refer to this ball as the "Earth."

The *Sun Segment* is a part of a circle representing a segment of the Sun, around which the Earth revolves. It can be enlarged by drawing out the arms at each extremity of the Segment.

The *Sun Pointer*, or *Needle*, is used to represent the direct rays of sunlight. It is movable, and can be pressed down to the surface of the Earth.

The *Moon Orbit Arm*, or *Moon Bar*, is used to carry the Moon around the Earth.

The smaller ball represents the Moon. The semicircle fastened to the poles of the Moon is used to show the *Circle of the Moon's Illumination*, as seen from the Earth.

The movable *Semicircle* fastened to the poles of the Earth, is used in finding the latitude and longitude of places, and as an astronomical meridian. I also call it the *Meridian Circle*.

The *Day and Night Circle* divides the surface of the Earth into two equal parts. We call this circle the *Sunrise and Sunset Circle*, the western half being the *Sunrise Circle*, and the eastern half the *Sunset Circle;* and it is sometimes called the *Circle of Illumination*.

The *Twilight Circle* is 18 degrees from the Day and Night Circle. The portion of the Earth's surface between these two circles is called the *Twilight Belt*.

The following illustration shows the Tellurian ready for any experiment that has reference to the midsummer day. You will notice that the calendar index points at the 21st of June. By taking hold of the orbit arm, and moving it so that the calendar index will move toward July, you will get the direction that the earth moves in her orbit. By stopping at any date you will have the earth in her relative position to the sun, ready for any experiment that you may wish to try, for any location on her surface. By tightening the screw at the north pole, the movable meridian will become fixed, and you can rotate the earth on her axis to bring the desired locality, or place, to any desired position without moving it.

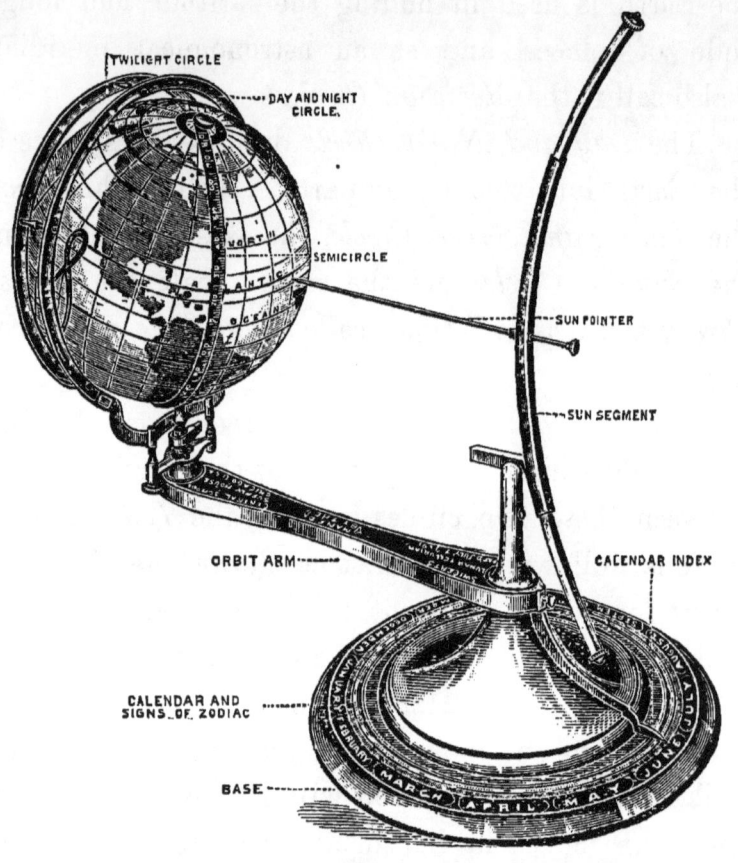

Place the movable meridian over the meridian of New York, then, by rotating the earth so that New York will move toward the sun pointer, you will get the direction that the earth rotates on her axis.

In the study of this Manual, it is expected that you will constantly refer to your Tellurian to prove the statements made, or to get a clearer idea of what is told.

CHAPTER II.

THE BEGINNING.

At the World's Fair, in the Art Gallery, was a painting entitled "The Young Astronomer." A country lad, sitting on a hillside watching the stars, occupied the foreground, while his humble home could be seen in the distance, far enough away not to disturb the reflections of the young thinker, whose face showed his appreciation of the sublimity of the scene upon which his gaze was fixed.

There have been many lads who thus have watched the heavens, and wondered why some of the stars seemed to move about during the year. The "Milky Way" could not have escaped their notice, and they have wondered at the meaning of this beautiful path in the sky. Sometimes a comet has made its appearance, and a new field of wonderment has been opened. Perhaps the moon has hidden the face of the sun, or has gone into the earth's shadow for a short time, and the young star-gazer has sought in vain for an explanation.

If the heavens could have been viewed with a

telescope, new wonders would have been seen on every side by these young astronomers. They would have seen beautiful white, fleecy clouds here and there, that are now thought to be the beginnings of solar systems like our own; and, if they could have understood the different objects in detail, they would have seen worlds in every stage of development, from Saturn's rings to our own place of abode.

But, while now we can definitely know but a few things concerning "other worlds than ours," still we can learn a great deal about our own that will be of interest and profit to us. We know that some rocks look as if they had been made in layers; and we know, that, if we could take muddy water and let it settle, a layer of dirt would be formed. If earth of another color should become mixed up with the water, and should settle, another layer would be formed at the bottom of the water. If these layers should, in the course of time, become hardened like stone, they would form rocks of different colors and materials, just as we have many times seen. From this we can tell pretty nearly how rocks that are found in layers have been formed.

Some rocks are glassy, and look as though they had been melted, which is probably just what has happened to them. From this we infer that the surface of the earth has at one time been at a much higher temperature than we can imagine it to have been. This carries us one step nearer to the starting point of the earth. Where there are very deep mines, it has been found that the temperature increases as we descend into the solid portions of the earth; and, by comparing different places, it has been found that it does not vary much from one degree for every fifty feet. At this rate of increase of temperature, the earth's crust would be at red heat at a depth of twelve miles; and, at a depth of one hundred miles, the temperature would be high enough to melt most of the materials of which it is composed. Now, a crust of one hundred miles in thickness is but a thin layer compared to the diameter of the earth, which is nearly 8,000 miles; and, if you could see a section of the earth, it would look something like the illustration here given.

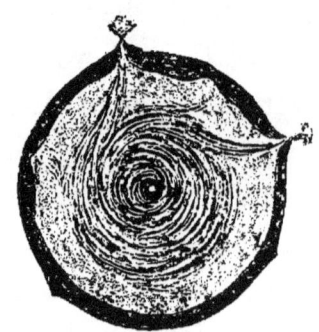

At some points the molten interior comes to the surface, being permitted to do so by great cracks in the earth's crust. In the course of time materials are piled up around these openings, making small mountains. Sometimes these openings occur at the tops of mountains, sometimes on level plains, and sometimes even in the bed of the ocean. They are called *volcanoes*.

You will find volcanoes on your Tellurian in South America, in Europe, and Asia,—in fact, in almost all of the grand divisions of the land and on some of the islands. Some islands are nothing more or less than volcanoes in the bottom of the ocean, that have piled up material around themselves until they have become islands.

It is thought that the earth, and all of the rest of the solar system, was at one time a cloud of nebulous matter, occupying all the space included in the orbit of our most distant planet; that this matter was intensely hot; and that, as it cooled, the particles came together, forming an immense ball that was one of the stars of the universe. As the mass revolved a portion was thrown off, and became what is now our most distant planet. As time passed on, another portion was thrown off, and formed another planet, and so on, until our earth and the remaining planets, in like manner, each began its separate course around the sun. The earth, in time, lost a portion of her matter, and from this the *moon* was formed, and began her revolution around the earth.

Just how the earth happened to get into the

position that she now occupies, no one can tell; but we know, that, if her position had been different, her conditions would also have been different from those that now exist, as you will see by the study of the following chapters in connection with your Tellurian, which represents the position of the earth in her relation to the sun, and is so arranged as to give you her movements as she performs her daily and yearly revolutions.

You will be able to see just how it happens that the days and nights are not always of the same length, and to see that they always would have been of the same length if the earth's axis had not been inclined toward the sun.

But before we attempt to study all the relations of the earth to the sun and the moon, we must learn something of the earth's surface; and the first thing to do is to learn the directions and the locations of all the most important places on the globe. This we will do in the next chapter.

CHAPTER III.

DIRECTIONS AND LOCATIONS.

There are a few points on the Tellurian that I wish you to learn, so that you can refer to them readily, for you will want to use them more or less in every lesson that you have to study with reference to locations and directions. The first two are the *North* and *South Pole*.

If you rotate the earth, you will notice that the cities near your own home will move around in a circle, and return to the point of starting. This they actually do once in twenty-four hours; and, if they did not, we would not have days and nights alternating with each other, but it would either be day or night all the time. The earth always rotates from the west toward the east; and, if you will place the earth so that your nearest city will be under the sunrise circle, and rotate it toward the sunset circle, it will move in an *easterly* direction, and it will be day to that city, until it reaches the sunset circle. If you continue the rotation in the same direction till the city reaches the sunrise circle again, it will be back at the starting point. While you are passing from the sunset circle to the sunrise circle, it will be night, and during this time the earth will still be revolving in an easterly direction, the same as during the day. It never revolves in any other direction. The opposite direction is *west*.

We started to learn how to find the north and the south pole on the Tellurian; but I wished to tell you how to find east and west first, because direction is determined by the earth's rotation. If you will find Greenland, and rotate the earth, you will notice that it makes a much smaller circle than your city did, for Greenland is farther north than your city, and, the farther north you go, the smaller the circles will become, until you reach their center, which is called the *North Pole*. All places that are between you and the north pole are *north* of you, and, if they are not in a direct line, they may be east or west of north from you.

If you go toward the equator from your nearest city, you will be going *south*. After crossing the equator, you will notice that, as you rotate the earth and watch the places during one revolution, the circles will get smaller the farther south you go till you come to their center, or the *South Pole*. All places in this direction will be south from you, and, if not in a direct line, they will be east or west of south from you. A line passing through the north and south pole, and also through the center of the earth, is called the earth's *Axis*, and the earth rotates upon this axis the same as a wheel does upon its axle. The rotation of the earth, then, determines the location of these two points.

The *east*, the direction of the rising sun, and the *west*, the direction of the setting sun, are probably the first cardinal points that were made use of by primitive man. They were used, without doubt, long before north and south were thought of. Probably men noticed that a certain star retained almost the same position in the heavens during each night in the year, and this afforded another reliable point for direction. All of the other stars seem to revolve around the star which is called the *North Star*.

This method of finding north and south, also shows you where the extremities of the earth's axis are, which are called *poles*, from a word that means a "hinge," a "pivot," or an "axis." So North Pole really means that end of the earth's axis that is toward the Pole Star. You can easily find the Pole Star, or North Star, on some clear night, if you will first find the "Great Dipper," and then look in the direction of the line in the cut below, which gives its position in the early evening of the 22d of September, or about that date.

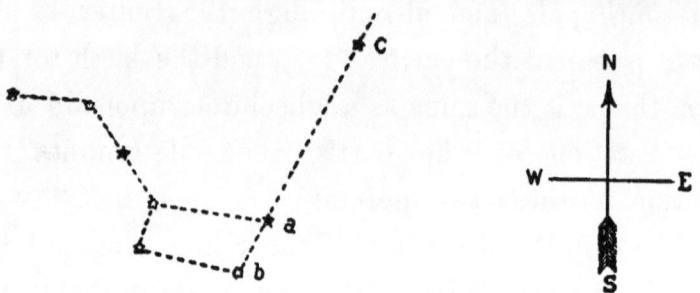

The Dipper lies with its handle a little above the bowl and pointing to the west. A line drawn through the two easternmost stars, *a* and *b*, will pass near the Pole Star, *c*. Two of the stars in the handle, and the star *a* in the bowl are much brighter than the other stars, and the Pole Star, or North Star, is brighter than any of the surrounding stars. If the axis of the earth could be extended to the heavens, it would pass near this star, and it is a strange fact that the Magnetic Needle indicates the same direction.

After the discovery of the properties of the loadstone, the invention of the *Compass* became possible, and all points of direction could now be determined without dependence upon the rising sun, the setting sun, or the North Star. Sailors were no longer obliged to wait for clear weather to find in what direction they were sailing, and became more venturesome, till they were at last able to cross the ocean without fear of losing their way, and to find distant points without being obliged to coast along the shore till they reached them. The Magnetic Needle always points toward the *Magnetic Pole*, which is near the North Pole, so near that it is called north, and all points of the compass are reckoned from the Magnetic Pole.

The following figure will give the principal points of direction, which you may use in giving the locations asked for in the last of this chapter.

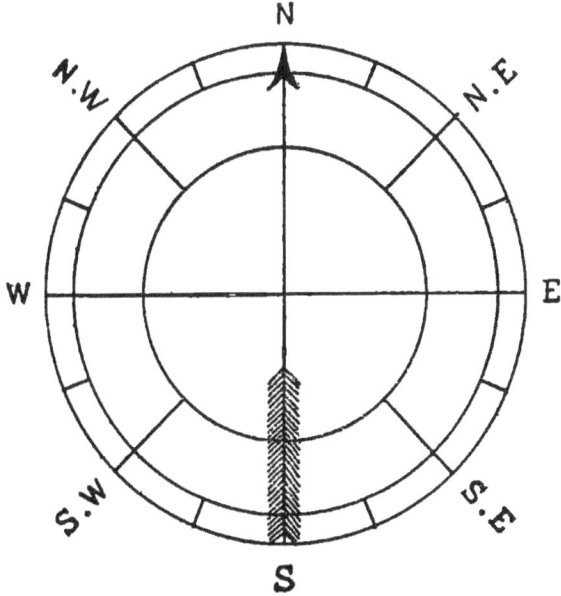

The point between N. and N. E. is marked N. N. E.,

and is read North Northeast; and between N. E. and E. is read East Northeast, etc. You will be able to read the other points opposite to these.

The other two points that I wish you to find are on the equator. One of them is where the line running north and south through London crosses the equator. It is marked with a zero. If you follow the equator just half way around the earth, you will find the other point. This is marked 180°. It is directly opposite the zero point. Places that are the direct opposites of each other on the earth's surface are called *antipodes*. This name is also applied to the people who live in the antipodes. Can you tell who are our antipodes?

These two points, with the first two mentioned, are important because they locate the principal lines that are used in giving the exact locations of all places on the earth. These locations are spoken of as the *Latitude* and the *Longitude* of places.

We sometimes speak of *Oriental* countries, meaning the countries of Eastern Asia. The East is sometimes called the *Orient;* but, when it is so called, the countries are referred to instead of the direction. When we speak of the *East* in this country, we usually refer to the New England States, and sometimes we refer to the Southern States as the *South*, and the Western States as the *West;* but, when we give the location of a place by giving directions, we usually select several well-known places, generally important cities, and tell in what direction it is from them.

Small islands, places at sea, and distant places are located by giving their latitude and longitude.

Direction has reference to the earth's surface only. We cannot say that the sun, the moon, or the stars are north, south, east, or west of each other or of the earth; but, when we refer to them, we compare their location with our own position and some other place on the earth's surface that is in the same direction. If we see the sun while looking southward, we say that the sun is south of us; and, if, in the early morning, we should look north of east and see the sun, we would say that the sun is farther north than we are; but we know that the sun can never get farther north than the summer solstice, which is about 23½ degrees north of the equator, and is farther south than the boundary line of the United States. It seems strange that the sun can shine into our north window when it first rises on a midsummer's day, and still be no farther north* than Havana, Cuba.

* Notice that the latitude of Havana is 23½° north.

In the study of the Tellurian, you will see why this is true; but there are a number of things to learn before investigating this point, and we will begin by studying the boundaries of some important countries on the surface of the globe.

We give the *boundaries* of any country, state, or body of water, by naming all the other countries, states, or bodies of water that touch it; and, in doing this, we usually begin at the north, then take the east, then the south, and then the west, naming all bodies of land or water that are contiguous. By giving the boundaries, we locate large tracts of land or water.

The following places will be used a number of times to illustrate experiments that will be made with the Tellurian, and it will be necessary that you should know just where they are. Locate them.

Karmakuli, St. Petersburg, Berlin, London (Greenwich is a part of this city, and is the location of the world's observatory, and the center of astronomical calculations), Chicago, New Orleans, Seattle, Quito, Valdivia, Wellington, the Antipodes Islands, the island near Enderby Land.

Mention five points for the class to locate.

LOCATION OF PLACES.

QUESTIONS.

Q. What is the location of Cuba?

A. Cuba is north of the most western portion of South America, east and a little north of Yucatan, and south of Florida.

Q. What are its boundaries?

A. It is surrounded by the waters of the Atlantic Ocean. The waters on the north are called the Florida Channel and the old Bahama Channel; on the east, the old Bahama Channel; on the south, the Windward Passage and Caribbean Sea; on the west, the Channel of Yucatan.

Locate England, and give its boundaries. Where is New Guinea? Give the boundaries of Australia. Locate Borneo, and give its boundaries. What are the boundaries of the United States? What are the boundaries of Europe? Locate Iceland, Hawaii, New Zealand, Tasmania, Sardinia, Graham's Land, Island of Ceylon.

Give the boundaries of Asia, South America, Greenland, India, Nova Zembla, Russia, Germany, Ecuador, Chili, and Enderby Land.

What are the points of the compass mentioned in this chapter?

Why are east and west given first?

What was the next point that was probably used? And what locates it?

What else have we that locates this same direction?

Do the stars of the Great Dipper seem to stand still?

What star do they help locate?

Tell how it is done.

What is the property of the loadstone that is mentioned?

What benefit comes from the loadstone?

What is the Orient?

How far north does the sun go?

Are we farther north than this?

Does the sun ever shine into the north window in this latitude?

Tell how we give the boundaries of any country.

CHAPTER IV.

MERIDIANS AND LONGITUDE.

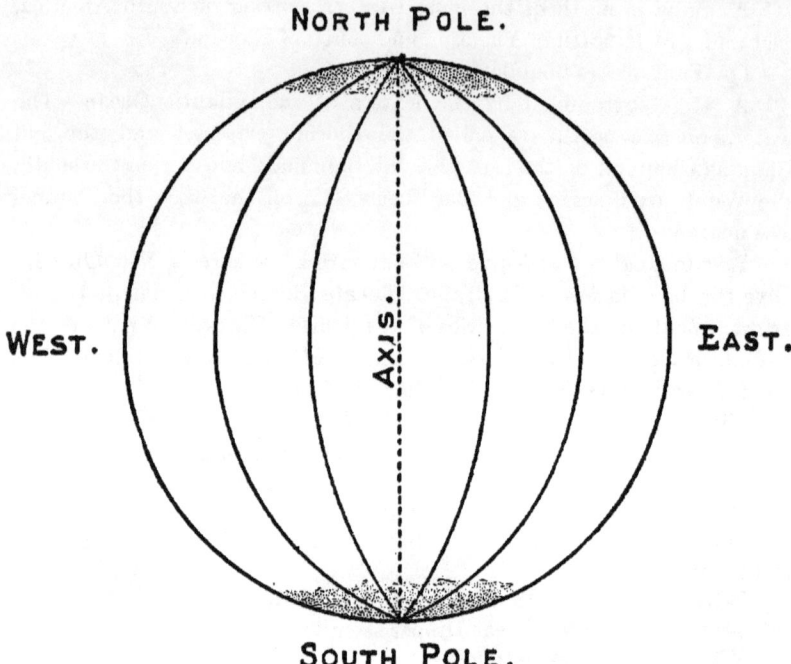

In the last lesson, we studied directions and locations; but, in giving the location of places, we were obliged to refer to some place near the locality that we were studying, and this place might not always be well known, even though it were important.

There is another way of locating that does not require the directions to be given, neither does it refer to any other place on the earth's surface. It is the method employed by sailors, and it refers to imaginary lines running north and south, and east and west, around the globe. They locate the place to which they refer, at their intersection; and then, by naming and numbering these lines, one may know exactly where the point is without regard to any other place. We will see if we can learn to locate places by this method also.

You will notice, on the surface of the globe, the two sets of lines that I have mentioned, and that they cross each other at right angles. One set runs north and south, and the other east and west. While but few of them are represented on the surface of the globe, still every place has these lines running through it, and they help give its exact location. These lines go clear around the earth, and form circles, which are divided into 360 parts, called *degrees*. Each degree is divided into 60 parts, called *minutes* (of distance), and these minutes are again divided into 60 parts, called *seconds*. One degree of a great circle is a little less than 70 miles, so one minute would be less than one and one-sixth miles,* and a second is less than one-fiftieth of a mile. So, by locating by means of degrees, minutes, and seconds, you get within less than one-fiftieth of a mile of the exact location; but, in the study of the globe, we will use only degrees and minutes in locating.

The set of lines that run north and south are called *Meridians*, and all meridians pass through the poles. If you put your fingers on the surface of the globe, and rotate it to the right (east), these meridians will pass under the needle that points toward the center of the earth; and, since this needle represents the direct rays of the sun, all places passing under it will have direct sunlight,—that is, the sun will be directly over them. When any point of a meridian is under the direct rays of the sun, all points on that meridian will have the noon hour. It is from the noon meridian that we begin to reckon the time of any place, and we say that all places on the noon meridian have 12 o'clock M. (M. is an abbreviation for *Meridies*.)

* The value of a single degree of longitude on the equator is equal to about 69⅓ miles.
    At latitude 45° it is equal to about 49 miles.
        "   60° "   "   "   35 "
        "   80° "   "   "   12 "
        "   90° "   "   "   0 "

Before we study the meridians in their relation to longitude, we will examine into their relation to time, for longitude is often determined by time. As I said above, all places have meridians, but not all of the meridians are represented on the globe. Only 24 of them are marked, and for convenience I shall call them *Hour-Circles*.* It requires 24 hours for the earth to make one revolution on her axis; and, if you divide 360 degrees into 24 equal parts, there will be 15 degrees in each part; so a difference of one hour in time will represent 15 degrees in distance, and the meridians that are given on the globe are 15 degrees apart, hence they have been given the name of "hour-circles."

These hour-circles on the globe always bear the same relation to each other, being geographical lines, located with reference to a line passing from the north pole to the south pole through Greenwich, England, and do not change their position with reference to places on the earth's surface; but, with

---

* "Hour-circles" is the name given to meridians that are 15 degrees apart on the Celestial Sphere, counting from the meridian of the sun, which is called the *Noon Meridian*. When we use the meridians on the globe for the purpose of estimating time, we take into consideration their distance apart only, disregarding the fact that the earth is in motion and that all of the meridians revolve with the earth.

reference to the direct rays of the sun, they move east at the rate of 15 degrees for each hour.

It will be 11 o'clock A. M. (*Ante Meridiem*) at all places on the first hour-circle west of the noon meridian, and 10 A. M. at all places on the second hour-circle west of the noon meridian. What will be the time at all places on the third hour-circle west of the noon meridian?

It is 1 o'clock at all places on the first hour-circle east of the one at which the needle points, and 2 o'clock at all places on the second east. What time is it at all places on the third hour-circle east of the one at which the needle points? Rotate your globe so that the needle will point to the zero meridian.

If you will look at England on the globe, you can find a city called London, with a meridian passing through it. This meridian is called the *Zero Meridian*, and, if you will follow this line down to the equator, you will find it numbered with a cipher. This is the meridian from which we reckon the position of distant places on the earth's surface. If you go east or west along the equator, you will find that the next meridian given, is marked 15, and the next 30. What will the next be marked?

If it is 12 o'clock at all places on the zero meridian,

what time will it be on the meridian 15 degrees west? What time on the meridian 30 degrees west? What time on the next hour-circle still farther west?

When we reckon the distance of any meridian east or west of the zero meridian, we call it finding its *longitude;* and, when we say that a place is 15 degrees *west longitude,* we mean that it is on the next hour-circle west of the zero meridian, and the clock will tell you that it is an hour earlier at all places on that meridian.* Can you tell what the longitude of the second hour-circle from this is? How many hours earlier is it on that hour-circle?

Follow the equator west, and see what is the greatest number of degrees any hour-circle has. All places between this meridian and the zero meridian are in west longitude. Find New York, and tell me in what longitude it is. In what longitude is San Francisco?

If you start at the zero meridian and go east to the first hour-circle, you will have gone over 15 degrees of *east longitude.* How many degrees of east longitude will it be when you reach the second hour-circle? What is the greatest number of degrees of east longitude that you can go?

* Remember, that, when we are talking about time, the 24 meridians shown on the globe are spoken of as "hour-circles."

If you can go to the next hour-circle beyond this, what longitude will it be? If you start at the zero meridian, and go east clear around the earth, what longitudes will you pass through?

In the last lesson we told you that the great circles divided the earth into two equal parts. We call each of these parts a *Hemisphere*, a word that means half of a sphere or globe. All of the meridians and the equator are *Great Circles*, and you can draw a great circle through any place if you will make it divide the earth into two equal portions. The equator divides the earth into a *Northern* and a *Southern Hemisphere*, and a great circle drawn through the 20th degree of west longitude and the 160th degree of east longitude, divides the earth into an *Eastern* and a *Western Hemisphere*. You see that they correspond nearly to east and west longitude. While all of Europe is in the eastern hemisphere, it is not all in east longitude, but a small portion of it is in west longitude. While the old geographers wished to reckon longitude from Greenwich, they thought it would be better to have all of the Eastern Continents in the eastern hemisphere, and this is probably the reason why they took the meridian of 20° as the dividing line. In studying the globe, it will be necessary for you to know the following table for the measure of circles:

60 seconds make 1 minute.
60 minutes " 1 degree.
360 degrees " 1 circle.

The Circle is sometimes divided into 12 parts called *Signs*.

The equator is divided into degrees; and, since there are 15 of these degrees between each of the hour-circles, each of these degrees is equal to four minutes of time, and half of a degree is equal to two minutes.

You can use the meridian circle to determine the number of the meridian that passes through any place, by rotating the earth so that the place will come under the meridian circle, and then, after seeing at what point it crosses the equator, counting the degrees to the first hour-circle east, if it is in west longitude, and the first hour-circle west, if it is in east longitude, and adding them to the number of that hour-circle. By practice you can divide the degrees into halves and quarters.

CHAPTER V.

PARALLELS AND LATITUDE.

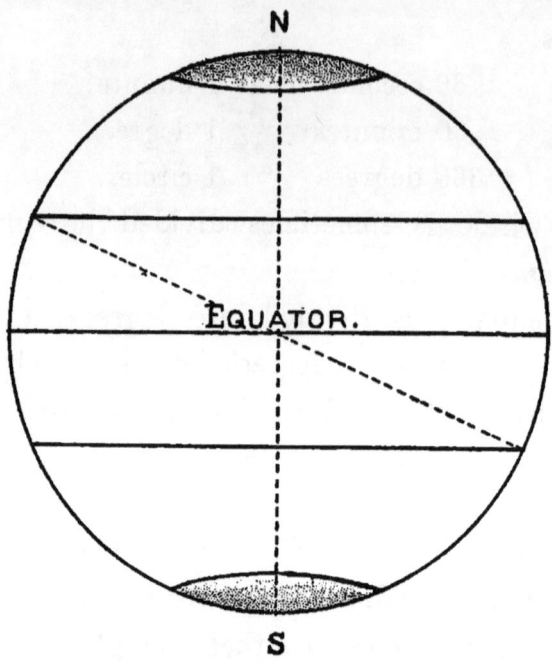

In the last chapter, I told you about the meridians, and said that the twenty-four shown on the globe were also called hour-circles. I told you that all places do not have the same time at the same moment, but that, when it is noon at one place, it is earlier at all places west of it, and later at all places east of it.

I also told you that there were two sets of lines on your Tellurian, and gave you the use of one of them, the meridians.

I will now tell you about the other, which you will be able to understand better by studying the foregoing figure. When we refer to distant places, we need to tell more than their longitude to locate them, and the lines I am going to tell you about in this lesson are used for that purpose. These lines go around the earth in the same direction that the equator does. They are called *Parallels*. All places on the same parallel are at the same distance from the equator.

Every place has a parallel passing through it, but all of the parallels are not represented on the Tellurian. Those given are ten degrees apart. In the figure at the beginning of this chapter, I have given only five parallels; but they are the most important on the earth's surface, for they do more than to help locate places. One of them, the *Equator*, is used as the basis of reckoning for all of the others; two of them, the *Tropics*, mark the limit of the sun's apparent movement north and south; and the remaining two, the *Polar Circles*, mark the largest area that remains, at any time during the year, in the day or night hemisphere for twenty-four hours or more. But this use of these lines will not be discussed in this chapter.

The parallels given are numbered; and, if you will find the 180th meridian, and go north or south of the equator, you will find these numbers, and they will tell you the number of degrees that the parallel is north or south of the equator. We call this finding their *Latitude;* and places north or south of the equator are said to be in *North* or *South Latitude*, and the equator, which is marked zero, is the dividing line.

The meridian circle on your Tellurian, is divided into degrees, and it is especially intended to be used in finding the latitude of any place. It has the advantage of remaining stationary while you rotate the earth so as to bring any place under it, and you can easily note the degree under which it passes; at the same time, by looking at the equator where the meridian circle crosses it, you can also determine the longitude of the place.

I find that the parallel of 20 degrees north latitude passes through the Sandwich Islands, and so we say, in locating them, that they are 20 degrees north latitude. By looking at the equator to see where the meridian circle crosses it, I find that they are 155 **degrees** west longitude.

LATITUDE.

Tell what countries the parallel of 40° north latitude passes through. What two large cities in the United States are near this parallel? Locate them.

If you go south of the equator, you would be in *south latitude*. Rio de Janeiro is about 23° south latitude. Can you find it? What is its longitude?

What states in the United States have the same latitude as Morocco in Africa? Can you give the latitude and longitude of the capital of the United States? In what latitude is Patagonia? Argentine Republic? What states in the United States are the same number of degrees distant from the equator as the Argentine Republic? In what latitude are they?

There are several things that affect the climate, but latitude is the most important. As a rule, the farther you go from the equator the colder it gets. This is one reason why we want to know the latitude of places, but there are a number of other reasons why we should study latitude. You will learn, in another lesson, why the days are longer in summer than they are in winter; but you must understand all about latitude first.

If you really understand all that I have said above, you can ask five questions about the latitude and longitude of some prominent places on the earth's surface, for your classmates to answer. If you cannot ask these questions and answer them without help, go over this chapter again, and then see if you can do it.

QUESTIONS ON CHAPTERS IV. AND V.

How many ways are there for locating places?
Which do you think is best? Why?
What are the straight lines on the globe called?
Are they really straight lines?
Into how many parts are they divided? What are the parts called?
About how many miles long is one of these parts?
About how many miles west of London is Philadelphia?
How far apart are the hour-circles?
Into how many parts must you divide an hour to tell the number of minutes that a degree is equal to?
In what direction from you is it earlier? Later? How many degrees from you is it two hours earlier? Three hours later?
Into what do the great circles divide the earth?
What is longitude? And what is its use? What is latitude? And what is its use? Does it have any other?
What is the use of the meridian circle that is on your Tellurian?
What places have parallels and meridians running through them?
What are the principal meridians? What are their uses?
Give some of the reasons why you should study latitude?
Which is the more important, a knowledge of latitude or of longitude? Why?
What is necessary to give in locating a place by means of lines?

CHAPTER VI.

STUDY OF THE GLOBE'S SURFACE.

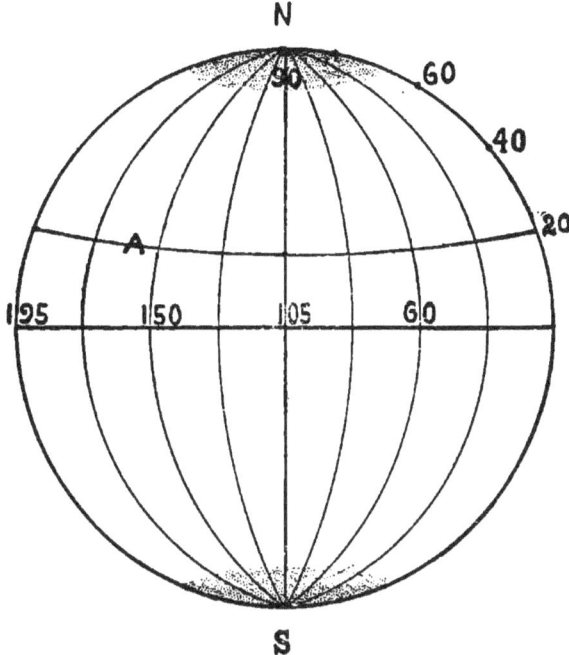

Above I have given you a figure that, I hope, will be of assistance in showing you how to use the Tellurian in the study of this lesson, and in the lessons that follow.

On the Tellurian is a movable meridian circle,

fastened at the poles of the globe. It has a set-screw at the north pole, so that you can fasten the meridian circle at any point that you choose, a very desirable thing to do when you are studying comparative latitude, or, as in the following lessons, in the study of time. You will notice that the meridians shown on the figure are an hour and a half apart, while those on your Tellurian are only one hour apart. I have numbered them on the equator, but the numbers that I have given the parallels are not the same as on the Tellurian. When they are numbered at all, they are numbered on the 180th meridian; but, on the Tellurian, it is not necessary to number them on the globe surface, for their numbers will be found on the meridian circle, which can be moved around the globe to any point, and fastened with the set-screw, or you can fasten it, and then rotate the globe till any point comes under the circle. By fixing the circle, and rotating the earth, you can easily tell all of the places that are in the same latitude, because they will pass under the same point on the circle.

The letter A in the figure is at about the location of the Sandwich Islands, and, in giving their location, you would say that they are exactly south of Alaska and west of Mexico; but, to be more exact, you would say that they are 155 degrees west longitude, and 20 degrees north latitude. Look it up and see if this is correct. You may be able to get it more exact than I have given it.

You will notice, on the surface of the globe, another set of lines that have not been mentioned, curving here and there through the oceans. They represent the *Ocean Currents*, and they are colored red to represent warm, and blue to represent cold currents of water which move continuously in the same direction. The names of the most important are given. These currents need to be studied because of their effect on climate; and, besides, they help determine ocean routes in navigation, especially for sailing vessels. In locating islands, it will be well to tell if they are in cold or warm currents of water.

I have selected the following points, on account of their favorable location, for comparison of length of days and nights, twilights and climate; viz., Karmakuli (Nova Zembla), St. Petersburg, Berlin, London (Greenwich is practically a part of London), Chicago, New Orleans, Seattle, Valdivia (Chili), Wellington (New Zealand), Antipodes Islands and Enderby Land. When I speak of Enderby Land, I shall refer to the island located near the point where the meridian of 60° east longitude crosses the antarctic circle.

Locate the above by giving latitude and longitude, using the meridian circle in determining both. Of course you can tell whether it is in north or south latitude, and whether in east or west longitude.

Give the latitude and longitude of New York, Quito, Tokio, Rio de Janeiro, Calcutta, and Cape Town.

What city in Europe has the same latitude as New York? What is its longitude? Name all the places on this parallel of latitude.

Give the latitude and longitude of the Hawaiian Islands. Where is Iceland? What is the latitude and longitude of the mouth of the Amazon River? What city has the same latitude? Name all other places in the same latitude.

How many degrees from the most eastern to the most western point of the United States and her territories? What part of the entire distance around the world is this? How many degrees from this most western point to Pekin? Are these points in the same longitude? What large island is partly in east and partly in west longitude? What island has the greatest east longitude? To what country does this island belong? The man who first sailed around the world stopped at this island, and found some very peculiar people. Can you tell any thing about them? Not far from here is another island that is important for its coaling stations, and the United States controls one of its ports. Can you name the island, and give its latitude and longitude?

Name all of the islands that have no latitude. Are there any that have no longitude? Are there any cities that have no longitude?

Ask five questions for your classmates to answer.

What ocean current is off the coast of England? Where does it come from? Where does it go? It sometimes washes trees ashore; what kind of trees do you think they are? Does it affect the climate of England? Which side of Iceland is the warmer? Why? Is Iceland warmer or colder than Greenland?

When warm and cold currents come together, they produce fogs. What parts of the ocean do you think are most liable to have fogs? The waters of the North Sea are cold, of the English Channel warm. What would be the result at London?

TEST QUESTIONS.

New Orleans has the same latitude as Coquimbo (Chili). Which is the warmer? Which is the warmer, the eastern or the western coast of the United States?

What two mountain ranges in the western part of the United States? What range in the eastern part? What ranges in South America? What is the general direction of these ranges?

Name and locate the ranges in Europe that are given on the globe. What is their general direction? While Europe and Asia are usually called two continents, they are really but one continent divided into two parts by a mountain range. What is it called? Do the mountains of Asia run in the same general direction as those of Europe? Name the most prominent range in Asia. They are the tallest mountains in the world, and Asia is the largest body of land.

In what direction is the greatest extension of North and South America? Europe? Asia? What comparison do you notice in regard to this and the length of the mountain ranges? What comparison do you notice in regard to the length of the coast lines and the mountain ranges? In what part of a continent do you think you would usually find the longest ranges of mountains? The tallest mountains?

Where are the desert regions? Where are the largest lakes? Inland seas? Where is the longest river? What is the latitude of its mouth? Why could there not be so long a river in Asia? What is the longest river in Africa?*

* Continue this line of work until the entire surface of the globe is familiar to the pupils.

CHAPTER VII.

LONGITUDE AND TIME.

In the last lesson, I told you that you would have to understand all about latitude and longitude before I could tell you why the days are longer in summer than they are in winter. Sailors are obliged to understand latitude and longitude for another reason. If they did not, they could not tell where they were when crossing the ocean, and might sail for months and never find land. After a storm they find where the winds have driven the ship, by finding the latitude and longitude of the point where they happen to be. Let me tell you what they do, and see if you can tell me in what direction the winds have been blowing. You cannot do this unless you can tell the direction of the hour-circles that have earlier or later time than the actual time given by the meridian.

If I should set my watch by the local (not the "Standard") time at Chicago, and travel to some distant place, I should still have Chicago time; and, if I should look at my watch to see what time it was, it would not give me the actual time of the place where I was stopping, but the time at Chicago for that moment; and, by seeing whether the actual time was earlier or later, I could tell whether I was east or west of Chicago. Suppose I should look at my watch shortly after noon at the place where I was stopping, and it told me that it was sixteen minutes past 2 o'clock, which would really be the time at that moment in Chicago. Can you tell me in what direction the place would be from Chicago? Now, if I should find the latitude, it would tell just where I was located. The city is about four degrees south of Chicago. What city is it?

If a ship should leave New York, the captain would carry New York time; and, when the ship crossed the sun's meridian, he would know that it was the noon hour by the true time. If he should look at his timepiece to see what time it was in New York at that moment, he could tell in what direction he had sailed, and how many degrees he had gone; so he could tell how many degrees east or west of New York the ship was.

Suppose that he should find that it was 9 o'clock A. M. by the ship's chronometer, for that is what they call the ship's clock, what is the direction? How many degrees is the ship from New York? If it is on the same parallel of latitude as Cuba, and it is the noon hour on the ship's meridian, and the ship's chronometer says that it is 5:30 P. M., New York time, about what is its latitude and longitude? Near what islands is it? Can it get back to New York without crossing the equator? Why?

In another lesson I will tell you how sailors find their latitude, and then you can see how they cross the ocean without losing their way.

There are two ways by which you can change longitude into time, and you will find that your Tellurian will do it much easier than your arithmetic. Let us find the difference in time between London and St. Petersburg. This will be one of the easiest problems that I could give you, for London is on the meridian of Greenwich, so you can begin at zero to count.

Fix the meridian circle so that London will be at its eastern edge, that is, so that the meridian of Greenwich will be just under the eastern edge of the meridian circle; then rotate the globe toward the west. You will pass over two hour-circles before you reach St. Petersburg, and you will find St. Petersburg about one-fourth of a degree east of the second hour-circle. By looking on the equator and counting the degrees, you will find fifteen between any two hour-circles; so one degree must be equal to one-fifteenth of an hour, or four minutes, and one-fourth of a degree must be equal to one minute. I would not try to get the time nearer than minutes. You can easily see that the difference of time between London and St. Petersburg is two hours and one minute. It would be better, however, to count from the city that is farthest east to the western city, for then you would be rotating the globe in the direction in which the earth revolves on her axis.

Let us find the difference in time between London and New York. London is the more eastern city, so it will be the proper place at which to begin to count. Do as you did before,—fix the meridian circle at the meridian of London, rotate the globe toward the east and count the hour circles as they pass under the meridian circle. You will notice that four hour-circles go under the meridian circle before you reach New York, and that New York will come to the meridian circle just before you reach the next hour-circle. Look at the equator, and count the degrees passed over between the last hour-circle and the meridian of New York. You will find that there are fourteen, and since each is equal to four minutes, the distance will equal fifty-six minutes; so the difference in time between London and New York is four hours and fifty-six minutes. Find the difference in time between London and Hawaii. How many degrees is it? Multiply this number by four, and what will it give? How many hours and minutes is it equal to?

Find the difference in time between New York and Hawaii. The following questions will assist you: How many degrees from New York to the first hour-circle west of New York? How many minutes in time to this hour-circle? How many hour-circles between New York and Hawaii? How many degrees between the first hour-circle east of Hawaii and Hawaii? How many minutes? Adding the hours and minutes that you have found, will make the number of hours and minutes that represent the difference in time between these places.

Find the difference in time between St. Petersburg and New York in the same way. How many degrees is it? What is the difference in time between San Francisco and Hawaii? What is the difference in degrees?

What is the difference in time between New York and San Francisco? What is the difference in degrees? When it is one o'clock P. M. in New York, what time is it in San Francisco? What time is it in London? The direction in which the earth rotates will tell you whether it is earlier or later. A telegram sent from New York to San Francisco reaches San Francisco at noon. It takes less than a minute; at what time was it sent?

- Yesterday's Liverpool markets are quoted in our morning papers. Suppose the markets close at 3 P. M., could our evening papers quote them? At what time would the cablegram reach Chicago if it was sent at once and there was no delay?

The following is a table of the longitudes of several places as given in the Gazetteer. It will be of assistance to you in making problems similar to those given, and will furnish you with proof of the correctness of your work.

Antipodes....................East longitude 178°, 43′.
Hawaii.......................West longitude 155°, 40′.
San Francisco................West longitude 122°, 24′.
New York.....................West longitude 74°, 00′.
London ....   00°, 00′.
Cape North...................East longitude 25°, 46′.
St. Petersburg...............East longitude 30°, 18′.
Apia.........................West longitude 171°, 21′.
Juan Fernandez...............West longitude 78°, 53′.

At the close of Chapter IV. you will find the table of circular measure, which is used in the study of arithmetic for finding longitude and time. The following table of comparisons will also be useful:

        360 degrees, — 24 hours.
        15 degrees, — 1 hour.
        1 degree, — 4 minutes.

Seconds of distance and of time need not be considered in the problems that follow, so they are not included in the preceding table. The diagram below will also assist in understanding the table, and the rule for the solution of the problems that are given:

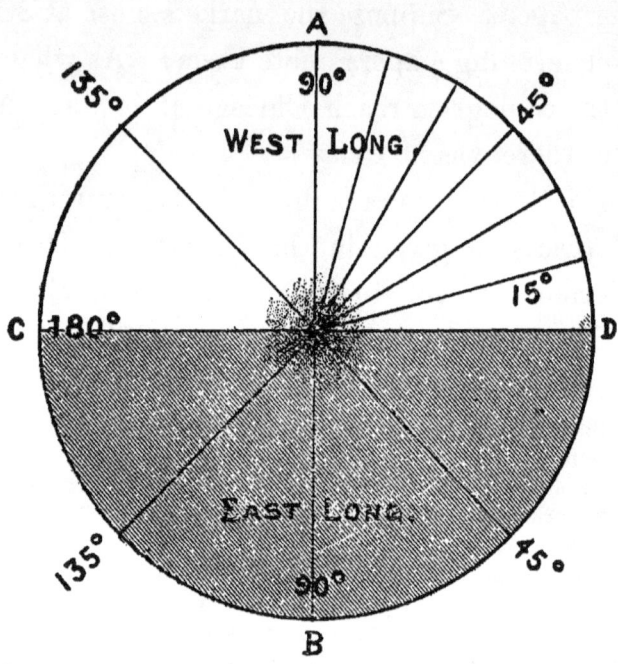

The circle represents the equator, and D is the longitude of Greenwich. The first 90 degrees of west longitude are divided into six parts, each equal to one hour, and each equal to fifteen degrees. The line C D is the dividing line between east and west longitude. To find the difference between 30° west longitude and 90° west longitude, you subtract.

DIFFERENCE IN TIME. 51

How many degrees is it? How many degrees is it from 30° west longitude to 45° east longitude? Did you find it by adding or subtracting? How many degrees is it from 90° west longitude to 135° west longitude? Did you add, or subtract, to find it? How many degrees is it from 185° west longitude to 135° east longitude? Did you add, or subtract, to find it? Do you add, or subtract, to find the difference when both are in west longitude? Do you add, or subtract, to find the difference when one is in west longitude and the other is in east longitude? When both are in east longitude how do you find the difference?

RULE.

I. Find the difference in longitude.

II. Divide this difference by 15, and the result will be the difference in hours. If you have a remainder, multiply it by 4, and the result will be the number of minutes.

1. When it is noon at San Francisco, what time is it at New York?

SOLUTION.

San Francisco is 122° west longitude.
New York is 74° west longitude.
Difference, 48°

15)48
3 and 3 remainder; that is, 3 hours and 3 times 4 minutes, or 12 minutes.

As New York is east of San Francisco, it would be later there than at San Francisco. Hence New York time would be 3h. and 12m. P. M.

2. What is the difference in time between Cape North and London? If it is noon at London, what time is it at Cape North? If it is noon at Cape North, what time is it at London? Try the same problem with the globe.

3. An explosion occurs at St. Petersburg at 8 A. M. The news is telegraphed, without loss of time, to the London and New York papers which go to press at 3 A. M. Can they print the account in the morning issues?

4. A sea captain having New York local time, looks at his chronometer when his ship's meridian is at the noon hour, and finds that it indicates 12 minutes after 7 P. M. He is in latitude 49° 42′ south. Near what islands is he?

Since each degree is equal to 4 minutes in time, the following rule affords an easy method of changing longitude to time:

RULE.

I. Find the difference of longitude in degrees between the given places.

II. Multiply this difference by 4. The product will equal the number of minutes in the difference of their times. Change the number of minutes to hours and minutes.

    5. Find the difference in time between New York and Apia.

        Apia is   171° 21′ west longitude.

        New York 74° 00′ west longitude.

        Difference 97° 21′

The 21 minutes is equal to a little more than one-third of a degree. Then $97\frac{1}{3} \times 4 = 389$, and 389 minutes equals 6 hours and 29 minutes, which is the difference in time between Apia and New York.

6. When it is noon at Wellington, what time is it at Berlin?

7. What is the difference in time between Melbourne and Chicago?

8. On the 23d day of June, the sun rises at 5 o'clock in New Orleans. What time is it then in London? What time is it then in Wellington? New Zealand?

CHAPTER VIII.

DAYS AND NIGHTS.

The sun is our principal source of light, and shines on half of the earth all the time. I have already told you that the earth is divided into hemispheres, and that the equator is the dividing line between the Northern and the Southern Hemisphere; also that two of the meridians divide it into an Eastern and a Western Hemisphere. I wonder if you can tell where those meridians are.

There are four more hemispheres that I want to tell you about, a *Land*, a *Water*, a *Day*, and a *Night Hemisphere*. If you draw a great circle around the globe so that it will pass through a point located at 45 degrees north latitude, and 150 degrees west longitude, and another point at 45 degrees south latitude and 30 degrees east longitude, a greater part of the land will be north of this line, and most of the water on the earth's surface will be south of this line. Can you find these points, and stretch a rubber band around the globe so as to show the Land and Water Hemisphere?

The day and night circle is a great circle, and, in using the globe, we call it the *Sunrise and Sunset Circle*. It divides the earth into two equal parts, which I have called the Day and Night Hemispheres. These hemispheres are changing their position all the time, and in this respect are unlike the Northern and Southern Hemispheres, whose boundaries are fixed by the equator, or the Eastern and Western Hemispheres, whose boundaries have been fixed by men who have made geography their special study. The Northern and Southern Hemispheres cannot change their positions, for they are separated by the equator, a fixed line midway between the poles, which are fixed points, being the ends of the earth's axis. This axis does not change its position except by a very small amount in thousands of years, so we may say that it is fixed.

Now, what has all of this to do with the Day and Night Hemispheres? I will tell you. I wanted to show you the difference between hemispheres that could not change their positions, and those that were constantly changing, like the Day and Night Hemispheres. The *Day Hemisphere* is the half of the earth that is toward the sun, and the opposite half is the *Night Hemisphere*. If you will rotate the globe, you will see that places are constantly passing under the sunset circle from day into night; and, when the rotation brings your own home under the sunset circle, you say that the sun is going down; still, you know that the sun does not move in this way, but only *appears* to move.

If you were riding on a train of cars, and should look out of the window at the fence-posts along the track, they would appear to be going in the direction opposite to that in which you were going, and the fence would appear to be actually moving, just the same as the sun appears to be moving when it is going down.

Rotate the globe until New York is under the sunrise circle, and it will represent the position of New York when the sun first appears in the eastern sky. As you continue to rotate the globe, it will illustrate New York's change of position, which causes the sun to appear to rise above the ground. If you watch the needle until the first hour-circle comes under it, it will tell just how long the sun has been up; but I will tell you all about this in the lesson on the "Length of Days and Nights." You will see, as you continue the rotation, that New York moves toward the noon meridian, passes it, moves on toward the sunset circle, reaching it just at sun-down, and passes into the night hemisphere.

It then moves toward the midnight meridian, passes it at 12 o'clock, and goes on toward the sunrise circle. When it reaches this circle, the earth has made one revolution on her axis, which has taken just twenty-four hours; and, no matter if the day is longer than the night, or the night longer than the day, both together have the same number of hours that it takes the earth to make her daily revolution.

So you see, that, while the day hemisphere is fixed in relation to the sun, it appears to go around the earth from east to west, constantly bringing places into the day hemisphere, and leaving in the night hemisphere the places it has passed over; but, really, it is the rotation of the earth that does this, and the day hemisphere does not move at all, but, like the fence-posts, only appears to move. Each place is on the side of the earth on which the sun shines, for a part of the twenty-four hours only; and, for the remainder of the twenty-four hours, it is on the dark side of the earth. Can you see how this is true?

In another lesson I will tell you why the rotation of the earth does not carry some places on its surface into the night hemisphere for several months at a time, and then for several months does not carry them out of it. For weeks they have long moonlight nights, with a few hours of twilight; but the sun does not appear. At last it just rises above the horizon, and in a few minutes sets. After this the days grow rapidly longer, until the sun does not go down at all for the remainder of the year. Day and night do not mean the same, to people who live in these places, as they do to us.

The moon has but one day and night for twenty-nine and one-half days and nights; so a day and a night on the moon are each a little more than two weeks long. If you watch the moon for one of its days and nights, you can see just how the days and nights on the earth would look if you could be on the moon and watch them, and see how the night appears to follow the day. You can see just how the day hemisphere looks, and that it is always on the side toward the sun. You can also see how it is, that, in like manner, the rotation of the earth actually carries places across the day and night hemisphere.

Begin to watch the moon when it first appears in the western sky, just after sun-down. You will see only a narrow strip of the day hemisphere on the western edge, a shining crescent, the new moon, made luminous by its reflection of the sunlight. Almost all of the night hemisphere is toward you, so that you do not see the moon as a round body, unless the earth's atmosphere is very clear, when a faint outline of the moon is made visible by the light reflected from the earth. If you watch the moon each night, you will see that the part that shines, grows larger for two weeks; in other words, the day hemisphere has in two weeks crossed over that part of the moon's surface that is toward the earth, until you see the whole of the day hemisphere, the full moon appearing in the eastern sky just after sun-down.

The next night it will look a little flattened on one side, and the next night still more so, until, at the end of the week, one-half of the moon will be bright, and the other half dark; that is, you will see half of the day hemisphere, — one-quarter of the moon, — and half of the night hemisphere. At the end of another week the whole of the night hemisphere, or the dark side, will be toward you, and you cannot see the moon at all; but, in a day or two, you will see it again in the western sky, just after sun-down, a new moon, just as you saw it twenty-nine days before, when you first began to watch it.

The bright part of the moon is always toward the sun; and, if you should draw a line from one point of the crescent to the other, and then another line through the middle of this line and the middle of the crescent, it would point directly at the sun.*

* See Chapter XIV.

CHAPTER IX.

YEARLY REVOLUTION OF THE EARTH.

So far you have studied only one revolution of the earth. It has another, its revolution around the sun, or *yearly revolution*, which causes the changes of the seasons, and the constant variation in the length of the days and nights.

Place the earth at the 20th* day of March, in its orbit, and press the sun pointer down close to its surface, then move the orbit bar so that the index approaches April, and you will notice that the sun pointer has moved from the equator to a little north of the equator. If you rotate the earth the sun pointer will point to the same parallel during the entire revolution. If you move the orbit bar so that the calendar index points at the 1st of May, you will notice that the sun pointer is still farther away from the equator, and that it will remain over the same parallel during an entire rotation of the globe. If you move the orbit bar again, so that the

* The exact time is not the same for different portions of the earth's surface; and, for the same portions, it varies each year, but within a small limit.

calendar index will show that the earth is at that point in her orbit which we call June 1st, you will see that the sun pointer is still farther north; and, when the earth reaches June 21st, the sun pointer will point very near to the parallel of $23\tfrac{1}{2}°$ north latitude. When the earth passes this point, the sun pointer will move back south again, and cross the equator on the 22d of September.

The direct rays of the sun continue southward till the 21st of December, when they have reached their southern limit, and the direct rays are at about $23\tfrac{1}{2}°$ south latitude, the southern tropic, or turning point, of the direct rays. Continuing the revolution of the earth around the sun, you will see that the direct rays travel northward, and reach the equator on the 20th day of March. This is called the yearly revolution of the earth, and it requires a little more than 365 days for its completion.

There is a figure on the globe called the *Analemma*, which gives the latitude of the direct rays of the sun for each day in the year, and sometimes it has a scale for the comparison of sun time with the time shown by a clock dial. You will find this figure on your globe just west of the northern part of South America, with the equator dividing it into two equal portions.

REVOLUTION OF THE EARTH.

Place the earth again at the 20th day of March, in its orbit, and rotate it so that the analemma will be under the direct rays of the sun; then move the earth around in the orbit, keeping the analemma under the direct rays of the sun all of the time that you are moving the orbit arm from March through each month till you get back to March again. Watch the sun pointer as it goes north, turns back south, crosses the equator, and goes to the southern point of the analemma, turns north again, and reaches the starting point in March, passing through all of the months of the year. This is an easy way to find the position of the earth in its orbit, and to get the latitude of the direct rays of the sun for any day in the year. It is important to consider the latitude of the sun's rays, when studying climate, and the length of the days and nights.

The earth's orbit has twelve divisions, called *Signs*, which correspond to the months of the year. The following table gives the name of each season, with the names of its signs, and of their corresponding months.

Spring..	Aries. Taurus, Gemini,	March. April. May.	Autumn	Libra, Scorpio, Sagittarius,	September. October. November.
Summer	Cancer, Leo, Virgo,	June. July. August.	Winter..	Capricornus, Aquarius, Pisces,	December. January. February.

THE SIGNS.

You have probably noticed the names of two of the preceding signs in your geography, in connection with the names of two parallels of latitude, the *Tropic of Cancer*, and the *Tropic of Capricorn*. "Tropic" comes from a word that means to turn. During the spring months the sun appears to come north until the earth has passed nearly through the sign of Cancer, and the direct or perpendicular rays have reached the Tropic of Cancer; then the sun seems to turn and go south until the earth is nearly through the sign of Capricorn, and then to turn again and go north for six months.

While at the turning point, it seems to stand still for a few days, and we say that the sun is at his *Solstice*. This word is derived from two words, *Sol*, which means "sun," and another word that means "to stand still." The solstice in June is called the *Summer Solstice*, and that in December the *Winter Solstice*. The point in the earth's orbit where the direct rays of the sun cross the equator when spring begins is called the *Vernal Equinox*, and the point where they cross at the beginning of autumn, the *Autumnal Equinox*. Can you give the dates when the earth is at these points?

This moving to and fro of the direct rays of the sun between the tropics is caused by the earth's yearly revolution around the sun. This revolution causes the changes of the seasons, and makes the days and nights of unequal length, except at the equinoxes, when all places on the earth's surface, save a small space at each pole, have equal days and nights. The word "equinox" is derived from *æquus*, meaning "equal," and *nox*, "night." All places on the equator have equal days and nights during the entire year; but the farther you go from the equator the greater will be their difference in length, except at the time of the equinoxes. Also the farther the earth is from its equinoxes, the greater this difference will be; and when it has reached the point in its orbit most distant from the equinox, this difference will be at its greatest. What are the points in the earth's orbit farthest from the equinoxes called? At what dates does the earth reach them?*

EXPERIMENTS.

1. Place the earth at the vernal equinox, and rotate. What do you notice regarding the latitude of the direct rays of the sun?

---

* Four seasons, 1894; central time: Vernal equinox, sun enters Aries, March 20th. Summer solstice, sun enters Cancer, June 21st. Autumnal equinox, sun enters Libra, September 22d. Winter solstice, sun enters Capricornus, December 21st.

EXPERIMENTS. 65

2. Move the orbit bar from the vernal equinox to the summer solstice. Is the northward movement of the direct rays uniform? If not, through what degrees of latitude do they pass most rapidly? How many days does it take for them to pass between these two points? How many degrees do they travel during the first thirty days? The second thirty days? During the remainder of the time? How many degrees do they pass through during the month of June?

3. Move the orbit bar from the summer solstice to the winter solstice. Tell what you notice regarding the rate at which the direct rays travel south, how long it takes them to reach the Tropic of Capricorn, and anything else you may have noticed.

4. Move the orbit bar back to the vernal equinox again. How many days has it taken to make the entire revolution? How many to go from the winter solstice to the vernal equinox? Is the period of time that the days and nights are nearly equal, greater or shorter than the time of their greatest difference? Why?

CHAPTER X.

LENGTH OF DAYS AND NIGHTS.

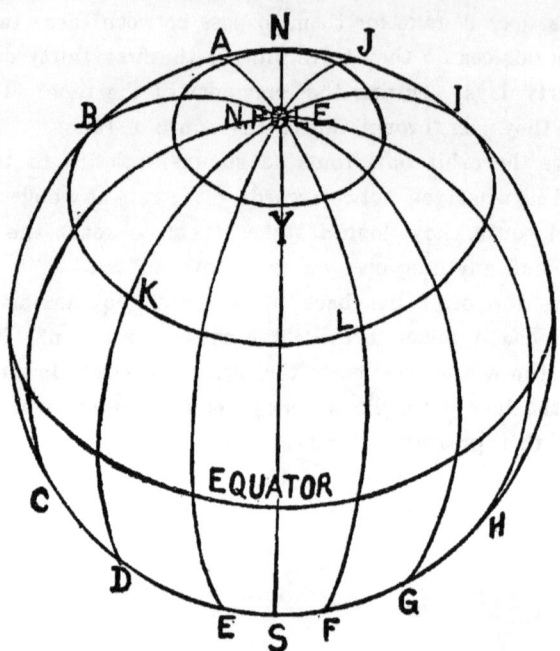

The length of the day at any place on the earth's surface depends upon the time it takes for that place to cross the day hemisphere; and the difference between the length of the day and twenty-four hours, is the length of the night for all places in the same latitude.

After the sun has crossed the equator, the entire length of that part of a meridian circle illuminated in any one day does not come into the sunlight at the same moment; but, if it is between the 20th of March and the 22d of September, the sun rises at the northern end of that part of the meridian circle before it does at the southern end, and sets at the southern end before it does at the northern.

The figure opposite represents the day hemisphere as it would appear on a midsummer's day, with the polar circle entirely within the circle of illumination, shown by the circumference or boundary line of the figure. It can also be seen that all the meridians cross at the pole. As the earth rotates on its axis, you see that no point within the polar circle is carried out of sight of the sun during an entire revolution. The point K has been in the sunlight since it left the point B, while the sun is just rising at the point D, which is directly south of K, for the same meridian passes through it. When D reaches G, the sun will set, but it will not set at K till K has reached I, which will be several hours after sundown at G. You can now refer to the Tellurian.

* See Chapter IX.

Place the earth at the 20th day of March, in its orbit, and, by rotating, you will see that the entire length of each meridian passes under the sunrise circle at the same moment; that is, every place has sunrise at the same time that other places exactly north or south of it have; also, that all places on the same meridian pass under the sunset circle and have sunset at the same moment, so that the day of each place is of the same length as the day at any other of these places; and this, of course, would also make the nights equal. They do not remain equal, however, as you will notice by moving the earth forward in its orbit; but the days will grow longer and the nights shorter at all places in north latitude, as the earth advances, until the summer solstice is reached. Let us see why this is true.

Press the sun pointer close to the surface of the globe, and measure the distance from its point to several points in the day and night circle, and you will see that it is at the same distance all the way round, and that the direct rays of the sun, represented by the sun pointer, are in the exact center of the day and night circle, or the day hemisphere. As the earth advances in its orbit the direct rays move north, and the entire day hemisphere moves north also. On the 15th day of April the direct rays will be at 10° north latitude; the southern limit of the day hemisphere will be just 10° north of the south pole, and all of the earth's surface north of 80° north latitude will be in the day hemisphere. The south pole will have no more sunshine for nearly six months, while for the same time the sun will shine on the north pole.

As you rotate the globe the northern part of each meridian will come into the day hemisphere before the southern part does, and the southern part will leave the day hemisphere before the northern part does; so the sun will rise earlier and set later at the northern part of the meridians than at the southern; that is, the days will be longer in northern than in southern latitudes. The farther north, the longer the day will be, and the longer the day the shorter the night. This is because one end of the earth's axis around which all the meridians rotate is within the day hemisphere, and it will be as far from this end (the north pole) to the nearest point of the day and night circle as the direct rays of light are north of the equator. Measure it, and see for yourself.

When the earth has reached the summer solstice the direct rays will be at about $23\frac{1}{2}°$ north latitude, the day hemisphere will include all of the arctic circle, and any one who is within the arctic circle can see the sun for the entire twenty-four hours. It will appear to travel around in a circle, and, if the position of observation is near the arctic circle, the sun will appear to nearly touch the horizon at midnight, and at noon to be a number of degrees above it. If the observer could be at the north pole, the sun would appear to describe a circle of which he would be the center, and, as the direct rays traveled toward the autumnal equinox, it would describe larger and larger circles each day, till finally it would make a circle just above the horizon all around, and in a day or two would go out of sight. Can you imagine how it would appear?

I have spoken a number of times about the *Noon Meridian*, and now I want to tell you about the meridian directly opposite, the *Midnight Meridian*. When the sun is at the summer solstice, the northern end of the midnight meridian, from the point where it crosses the arctic circle to the north pole, is in the day hemisphere, the sun can be seen from all points between these limits, and at midnight, in looking at the sun, you would look toward the north pole. It is called the *Midnight Sun*, and is always seen directly north of the observer. How can this be, when the sun never gets farther north than the Tropic of Cancer? Your Tellurian will tell you.

When the meridian of 75° west longitude is the noon meridian, what meridian is the midnight meridian? When the direct rays are at 20° north latitude, how far south of the north pole can the observer go and still see the midnight sun? In what direction will he look? Is the sun really north of the north pole?

So far we have confined our observations to the northern hemisphere during the spring and summer months. Let us see how it is with the southern hemisphere for the same time. Let us begin with our vernal equinox again;—and I say *our* vernal equinox, for it is the *autumnal* equinox to the people who live in the southern hemisphere, as will be shown in the chapter on "Climate." At this time the southern hemisphere, like the northern, has equal days and nights, for the entire meridian passes under the sunrise or sunset circle at the same time; but, as soon as the sun passes north of the equator, the southern hemisphere seems to turn away from the sun, and the south pole does not emerge from the night hemisphere for six months. The southern ends of the meridians pass quickly across the day hemisphere, and continue for a long time in the night hemisphere; and in length the days and nights are the reverse of what they are at the same time in the northern hemisphere. New Zealand has about the same latitude south that Illinois and Wisconsin have north, and the nights of New Zealand will be of the same length as that of the days in Illinois and Wisconsin at the same time; so that, while the latter are having long days and short nights, the former is having long nights and short days.

When the sun reaches his summer solstice in the northern hemisphere, the people in the southern hemisphere say that it is at the winter solstice; for they are having winter there, and their days are short, and their nights long. If you place the earth at the 21st day of June in its orbit, and rotate it, you will see that this is true.

In a former lesson I told you that the rotation of the earth carried all places on its surface around in a circle, and you have had an opportunity to see that it is true. Now I want you to notice another thing, which will tell you why the days and nights are not always of the same length; and which will be the most noticeable at the time of the summer solstice, when all places within the arctic circle make the entire circle in the day hemisphere, and a place at 70° north latitude makes nearly all of the circle in the day hemisphere. As you go farther south, these parts of a circle, or *arcs*, as they are called, grow more nearly equal until you reach the equator, where the day and night circle divides it into two equal parts, and here the days and nights are equal. But at 40° south latitude, the day and night circle divides the parallel into two unequal arcs, the longer arc being in the night hemisphere, and the night is longer than the day. By finding the number of degrees in these arcs you can easily determine the length of the days or nights by changing the degrees to minutes. There will be four minutes for each degree.

Study carefully the following rule for finding the length of any day or night, and, by using your Tellurian, show that the statements made are true:

RULE.

I. Move the orbit arm of the Tellurian so that the indicator will point to the 21st day of June. The polar circle will then just touch the day and night circle.

II. Rotate the globe so that Chicago will be under the edge of the sunrise circle. Without moving the globe, fix the meridian circle so that any convenient hour-circle will be just under its eastern edge. (In speaking of any of the circles the eastern edge is meant.)

III. Rotate the globe toward the east, and count the hour-circles as they come to the eastern edge of the meridian circle. Continue the rotation until Chicago comes to the night circle. The number of hour-circles that pass under the meridian circle will be the number of hours, and the number of degrees between the last hour-circle and the eastern edge of the meridian circle, when multiplied by four, the number of remaining minutes, in the length of the day.

IV. The difference between the length of the day and twenty-four hours, will be the length of the night.

Perhaps it may be well to leave the study and the application of the above rule until you have completed the next chapter.

CHAPTER XI.

(1.) HOW TO COMPUTE THE LENGTH OF DAYS AND NIGHTS.
(2.) TWILIGHT.

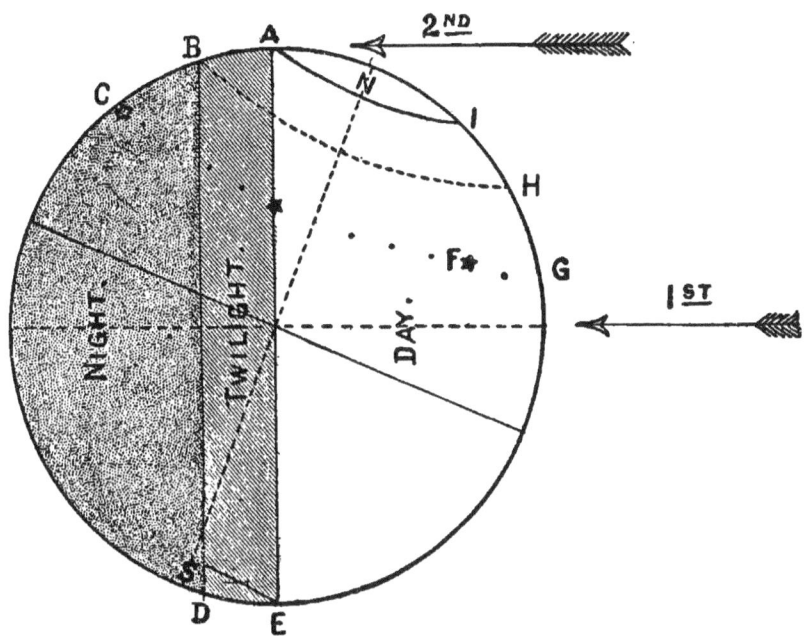

In the figure above I have represented the earth as it would appear to an observer if he could be situated so as to look at it from a station where the day and night circle, A E, would appear as a diameter, or where it would appear the same as the moon does when in quadrature. The dotted line NS is the earth's axis, around which all places on the earth's surface rotate once in twenty-four hours; the arrows represent the direction of the rays of sunlight; the light portion represents one half of the day hemisphere, or from sunrise till noon, and the dark portion, one half of the night hemisphere, or from midnight till sunrise, and includes the twilight belt. It is on the 21st day of June, the sun has reached the summer solstice, and, in the northern hemisphere, the days are longer and the nights shorter than at any other time during the year. The first arrow represents the direct rays of the sun, which strike the earth at the Tropic of Cancer; the second arrow, a ray of sunlight that strikes the earth at A, which is the greatest distance from the direct rays that it is possible for sunlight to touch the earth.

Since all places between A and N rotate about N, they will be in the sunlight during the entire revolution; but, when the direct rays move south, the point A will approach N, and the circle of continuous sunlight will grow smaller till A has reached N, which it will do on the 22d of September. (See figure on page 82.) When the point A is at N, the sunrise and sunset circle will coincide with two of the earth's meridians that are directly opposite to each other. The sun will be rising at every point in one of them, and setting at every point in the other. Two other meridians constitute the circumference of the figure; one of which is the noon meridian, and the other the midnight meridian. Name the points through which they pass.

Let us study the earth in the two positions mentioned,—at the summer solstice, represented in the figure, and at the autumnal equinox (see page 82), where the earth will be when the point A is at N,—and see if we can learn how to compute the length of days and nights, and then we can use the same method for the computation of days, nights, and twilights for any time during the year.

We will begin with the first position, with the earth at that point in her orbit called the summer solstice, and suppose that the city of New Orleans is located at the star C, a point in the midnight meridian. The rotation of the earth will carry it in the direction of the curved line CG. The dots along this line show the points at which it will arrive each hour from midnight till noon. It enters the twilight belt about 3 o'clock, and requires nearly two hours to pass through it and come in sight of the sun, which it does one minute before 5 o'clock, the time of sunrise at New Orleans for this date. At 6 o'clock it has reached the meridian NS, and at 10 o'clock, the star

F. If you now measure its distance from the sunrise circle, you will find that it is nearly the same as the distance from the midnight meridian to the sunrise circle. It is still two hours till noon, so the day at New Orleans for this time of the year is a little more than four hours longer than the night.

Let us follow this through with the Tellurian, and then you can try the same experiment with Chicago, and see if you can tell the length of the day and night, the time of sunrise, and the difference in the length of the day and night. Keep the earth at the same point in its orbit, note carefully each step as it is taken, and then you will make no failure.

Place the earth at the 21st day of June, which is its position in the figure at the beginning of this chapter, press the sun pointer down close to its surface, rotate till the meridian of Greenwich comes directly underneath the pointer, and then fix the meridian circle at the meridian of 180 degrees. The pointer will now serve the purpose of the sun's meridian, and the meridian circle that of the midnight meridian, so that, as the earth rotates, these two points will remain fixed, and you can count the hour-circles as they pass under them. In this way we can get the difference in time without getting the difference in longitude as the arithmetics teach.

Now, to find the time it takes for a place to pass from one point to another, as from sunrise to noon, rotate the globe till the place that you are observing is at the starting point. With the globe at rest in this position, find the number of degrees, if any, from the eastern edge of the meridian circle to the first hour-circle west of it, and multiply it by four to get the number of minutes; then, as you rotate the globe toward the east, count the hour-circles (after the first) that pass the meridian circle till the place has reached the given point, to get the number of hours. Then count the remaining degrees, if any, between the last hour-circle and the meridian circle, and this amount, together with the number of hour spaces, will give you the time required for the place to pass to the given point.

In this way let us count the time it takes for New Orleans to pass from the midnight meridian to the twilight belt, and then to the sunrise circle, which it will reach at one minute before 5 o'clock. Continue the rotation till New Orleans comes to the sun's meridian, or noon, counting as you go, and you will find that it will be seven hours and one minute till noon.

You may now try the same experiment with Chicago, which you will notice is a short distance from the first hour-circle west of it. You will also notice that the sun rises earlier, that the night is shorter, and that the day is longer. Be careful to see where the meridian crosses the equator, and remember that each degree is four minutes of time, and that 15 degrees is one hour of time. After you have tried this experiment, see if you can compute the time of sunrise at Chicago on April 25th.

In finding the length of the day, it is easier to move the meridian circle to the position of the sun's meridian, and then count the degrees* and hours that pass under it while the place that you are experimenting with rotates from sunrise to sunset. This time subtracted from 24 hours gives the length of the night. In the same way you can determine the length of twilight, by counting the degrees that pass under the meridian circle while a place is passing through the twilight belt.

Owing to the elliptical form of the earth's orbit and its inclination to the equator, the sun's apparent motion through the sky is not uniform; so the days are not exactly 24 hours long by the clock, but vary a little day by day, till the difference amounts on the 2d of November to over 16 minutes. A day of sun time is the time that it takes for a point on the earth's surface to rotate from the sun's meridian clear around

---

* Count each degree as four minutes, the spaces between the hour-circles as hours.

to the sun's meridian again; while a day of mean time (the time kept by a clock or watch) is 24 hours.

The time that you find with the Tellurian is sun time; and, to make it agree with the clock, it is necessary to correct it by adding to it, or subtracting from it, the difference between the sun's apparent time and the mean time, which is the sun's average time, compared with which the sun's apparent time is either slow or fast, according to the position of the earth in its orbit.

Sun time and mean time agree four times a year, on April 15th, June 14th, September 1st, and December 24th, and their difference is never more than 16 minutes and 20 seconds sun fast, or 14 minutes and 30 seconds sun slow, which occurs on November 2d and February 11th. This difference is called the Equation of Time, and will be found in any good almanac, for each day of the year, in a column headed "sun fast" or "sun slow." With this correction, you can do quite accurate work in finding time with the Tellurian.*

The following figure represents the earth at the time of the equinoxes, the second of the two positions previously mentioned. The line AE of the first figure coincides with NS in this figure, while A and B of this occupy the position of A and B in the first.

* See "Appendix," lesson on the Analemma.

DAYS AND NIGHTS.

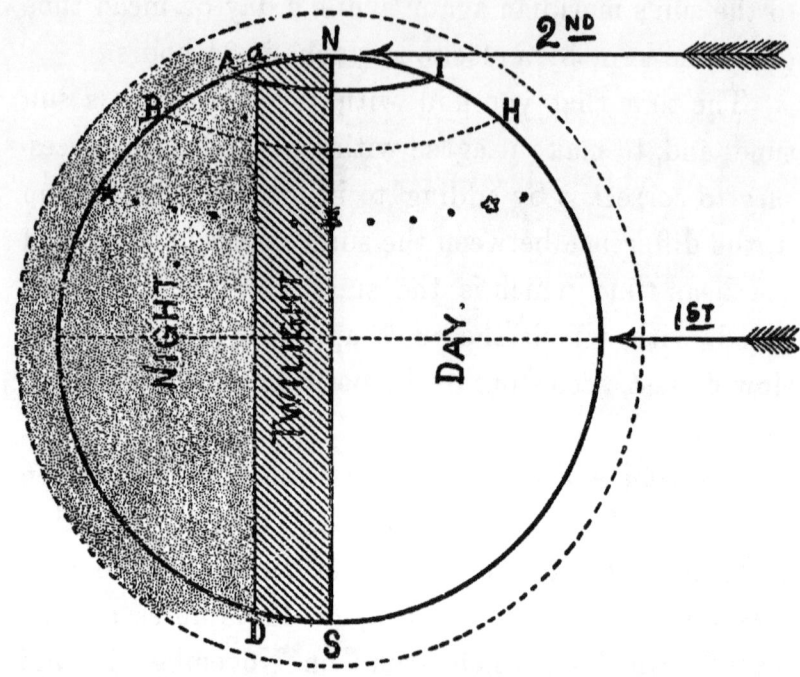

New Orleans is again at midnight, and the dots along the curved line show where it will be at the end of each hour from midnight till noon. Try the same experiments that you tried before, and note the differences in results. The point A is at the same distance from the pole that it was on the 21st of June; but see what a change has taken place in regard to the length of its days. How will it be ninety days later? Try the experiment, and see.

In this latter figure you will notice a dotted curved line surrounding the earth. This represents a stratum of the atmosphere 50 miles in height, the greatest height at which it is capable of reflecting sunlight. The sunlight in this upper stratum of the atmosphere can be seen until the sun is about 18 degrees below the horizon. This reflected light produces the phenomena of twilight.

If it were not for the twilight, it would become dark as soon as the sun went down; but, as it is, the first moments of twilight give nearly as strong a light as the last moments of sunlight. Each moment after sunset the rays of light pass through the atmosphere farther above our heads; and the farther they are above our heads the less light there is to be reflected, until, when the sun is about 18 degrees below the horizon, the last rays above us disappear and twilight ceases. This would occur in one hour and a quarter if the point of observation passed straight across the twilight belt; but all places pass through the twilight belt in a curve that is inclined to it. This inclination varies for different seasons of the year and for different latitudes. Many places stay in the twilight belt, at certain seasons of the year, during the entire night. Find some of them.

To get the length of the twilight of any place for any time, place the earth at the desired date in its orbit, then rotate till the place comes to the sunset circle, fix the meridian circle at any hour-circle, then count the degrees that pass under it while the place is passing through the twilight belt.

Now study the rule at the end of Chapter X. very carefully, so that you may be able to understand the reason for each step. In computing the time of sunrise for any place, it will be best to fix the meridian circle at the midnight meridian, then bring the given point under the meridian circle, and count the time from midnight till sunrise. To find the time of sun-down, begin with the noon meridian. To find the time between two given points, see the chapter on "Longitude and Time."

CHAPTER XII.

DISTRIBUTION OF LIGHT AND HEAT.

In Chapter VIII I told you that the sun was the source of light; it is also the source of heat; and since the sun shines on one-half of the earth all of the time, the heat must be greater on that portion of the earth's surface than on the part that is in the shadow; but all portions of the earth's surface that are in the sunlight do not get the same amount of light and heat. Let us inquire into the reason of this.

I have spoken a number of times about the direct rays of sunlight, and you might have inferred that there were other than direct rays; but this would not be strictly true, for the rays of light, or heat, as they come from the sun are parallel to each other, or nearly so, near enough to assume that they are parallel, so that, if the earth's surface were a disc perpendicular to the sun's rays, all places would receive the same amount of light and heat, and would have the same temperature. The curvature of the earth's surface prevents all places, except those in the center of the day hemisphere, from receiving rays of heat and light that are perpendicular; and it is the perpendicular rays that are meant by the "direct rays," as they are called in this book.

The curvature of the earth's surface causes a very unequal distribution of light and heat, as you will notice by examining the following figure. It will also show you the importance of the direct rays, and tell

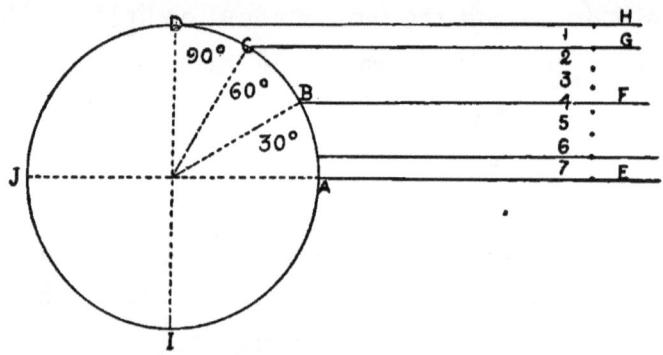

you why it is so much colder at the poles than it is at the equator. Notice, also, that the direct rays are the only ones that can be perpendicular to the earth's surface.

The figure represents the earth at the time of the equinox, and the dotted line AJ is the equator. The lines E, F, G, and H, representing rays of light and heat, are parallel to each other, and strike the surface of the earth at A, B, C, and D, which are at equal distances apart; EF and FH are the same distances apart, and from G to H is one-seventh of the distance from E to H. The figures represent equal distances.

If we could consider light and heat capable of being measured by extension, all of the light and heat that would fall on the surface of the earth between A and B would be equal to the light and heat that would fall on the surface between B and D; but from B to D is twice as far as from A to B; hence the light and heat, being spread over so much more surface, would lose a considerable portion of their power, and their direct effects could not be felt at all at the point D.

Study the figure carefully, and you will see that the light and heat that the earth's surface receives are the most intense under the direct rays, and that their intensity rapidly decreases as you go north or south from the equator. If you compare the different areas, you will find that the 30 degrees around the pole receive but one-seventh of the light and heat that the northern hemisphere does. You can see what a wonderful effect "the moving to and fro of the direct rays of the sun between the tropics" has on the distribution of light and heat. You can also see why it is so much colder in northern or southern latitudes than it is in the tropical regions, where the rays of the sun are more nearly perpendicular to the surface of the earth.

The same is true as you go east or west of the sun's meridian. This explains why the mornings and evenings are so much cooler than mid-day; and, if it were not for the rotation of the earth, a greater portion of its surface would become frigid, and we cannot imagine the intense cold of that portion directly opposite to the sun, while it would be equally difficult to tell what the effect under his direct rays would be.

The slope of portions of the surface of the earth also affects the distribution of light and heat the same as its curvature does; so that large sections of land that have rivers running to the northward do not receive the same amount of light and heat as equal areas in the same latitude having rivers running toward the south.

You have probably noticed that the snow on the south side of a roof begins to melt before the snow on the north side does. It is not because the sun does not shine upon both sides of the roof, but because both sides do not receive the same amount of sunlight, as the diagram on the following page will show.

The slopes AB and BC represent the same extent of surface to be warmed by the sun, and the distances between the arrows A, B, and C, represent the relative amounts of heat that each side will receive. You can readily see that the southern slope will receive double that of the northern slope; but, when the sun moves northward, so that the rays will be in the direction of the dotted line DC, the amount of light and heat that strike the northern slope will be increased; and, if the sun should get into a position so that the point B would receive the direct rays, both sides would receive the same amount.

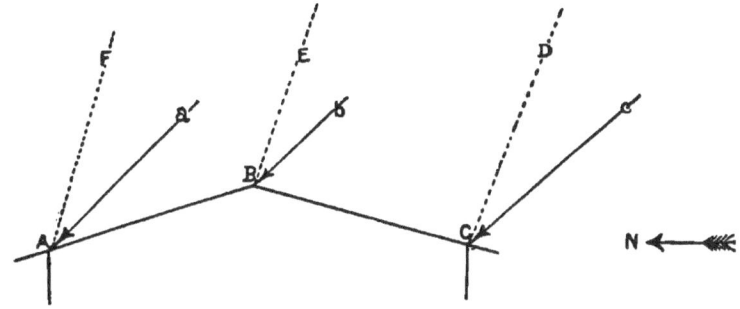

In studying the globe, you will see that some of the countries have mountain ranges running east and west, and you can see that they will have plains on either side whose slopes will correspond to the diagram we have just studied; and you can also see which side will be the warmer and why.

Thus there are five things to be considered in studying the distribution of light and heat: (1) the distance of any place north or south from the equator; (2) the yearly revolution of the sun, which carries the direct rays north and south; (3) the rotation of the earth; (4) the slope of the land; and (5) last, and most important of all, the winds and the ocean currents.

In applying these principles to the study of the globe, it will be well to begin with the earth at one of the equinoxes, and then you can make use of the parallels of latitude in measuring distances and comparing areas; but, when the earth is at any other point of her orbit, you must keep in mind the direct rays of the sun. You must also remember to take into consideration the mountain chains and the river systems, as they will indicate the slope of the land, which will modify all other things to be considered. Begin at the poles with the earth in the position shown by the figure near the beginning of this chapter. The sun can just be seen at the horizon, and the rays of light are parallel to the surface of the earth, so that they do not make any angle at all, in other words, do not seem to touch the earth; but, as you move a little farther to the south, lines drawn from the sun to the earth's surface make a very sharp angle; and, if you draw a line from the poles to the sun, and another parallel to it, from the 60th degree of latitude to the sun, they will not be very far apart; but, if you will draw another parallel line from the 30th degree to the sun, it will be three times as far from the line drawn from the 60th degree, as the line drawn from the 60th is from the line drawn from the 90th degree, and the angles formed with the surface of the earth will rapidly increase in size till you come to the equator, where the angle is 90 degrees. So, if 90 represents the maximum amount of light and heat received from the sun at the equator, it will be *zero* at the poles, and, for any place between the equator and the poles, it will be some number between zero and 90 that will correctly represent the inclination of the sun's rays.

It is a principle, that, if the temperature of air and water be increased, they will occupy more space, and have a greater power to resist the force of gravity. Hence they have a tendency to recede from the center of the earth, and, being perfectly mobile, they seek a position where this tendency is balanced by gravity. If the temperature of either be again increased, its position is again disturbed, and the same result follows. Of course, when these particles of air and water leave their original position, others take their places, and, in turn, become warmer, and give place for still others; so these elements can never be at rest, always having a tendency to follow each other, and currents of air or water are produced. As these currents carry with them the temperature of the places from whence they came, and gradually lose it as they come in contact with colder bodies, they become a distributing agency;

and warm winds may blow, or ocean currents carry, some of the warmth of southern climes to northern latitudes, as in the case of the Gulf Stream, which has such a wonderful effect on the temperature of the coasts of England and Norway. So a change of wind usually means a change of temperature, and, if we lived on the seacoast, the ocean currents would bring to us the temperature of some distant point. These are important facts in the study of climate, which will be the subject of our next chapter.

CHAPTER XIII.

CLIMATE.

The subject of climate is so closely related to the facts noted in the last chapter, that it merits some attention here, and the surface of a globe furnishes a splendid opportunity to study the theory of climate. By applying the facts in the last chapter, you can tell something of what the climate of any place is without knowing definitely beforehand.

If it were not for the part that the winds and ocean currents take in distributing heat, the location of any place would determine its climate, for its temperature would then depend upon conditions that are constant and uniform. To better understand this subject, it will be well to keep in mind the principal agencies that cause the distribution of light and heat, mentioned in the last chapter. Carefully determine which are constant, and which are not. There are variable conditions that will be considered here.

If you should warm a large stone, it would take some time for it to acquire the same temperature as the heat applied, and it would remain hot for some time after the source from which it derived its heat had been removed. It would then gradually cool until it was of the same temperature as its surroundings; and, the longer it took for it to absorb the heat that it had acquired, the longer it would take for it to part with it, and the more uniform would be the temperature of all contiguous objects, for, while cooling, it would impart its warmth to them. Also, the more rapid the diffusion of heat the greater will be the extremes of heat and cold when the source of the heat received is not constant, as in the case of the heat received by the earth from the sun. It is evident that the length of the days and nights is an important factor to be considered in this connection.

The diffusion of heat is dependent upon the law of nature, that all bodies, in the presence of warmer bodies, will absorb a part of their heat, and, in the presence of colder bodies, will give up a part of their own heat. This greatly modifies other natural conditions; and, as the tendency is constantly in the direction of maintaining an equilibrium, one change produces another until the whole is balanced, if that be possible.

The oceans and the continents cannot be affected to the same extent by the sun's rays, because of the mobility of the particles of water, and the immobility of the particles of which the earth's crust is composed. Then, too, these particles differ in kind, and some kinds receive heat more readily than others. Sand and muck exposed to the sun for the same length of time do not appear to be equally warm; for such badly conducting substances as sand convey the heat produced by the sun's rays downward, into the soil, with extreme slowness; hence it must remain longer on the surface, and in immediate contact with the air. This not only makes the surface of the ground much warmer during the day, but it permits the air passing over it during the night to rapidly absorb its heat. It is in accordance with this fact "that the climate of sandy deserts is characterized by nights of comparatively great cold." So, different kinds of soil have something to do with the climate; for, where the soil gives up in a few hours all the heat it has acquired during the day, the temperature becomes much lower during the latter part of the night; but, where the soil retains a part of its heat all night, the days and nights are more uniform in temperature.

The heat that the earth receives from the sun does not penetrate below its surface to any great extent, only a few feet, even in tropical countries; and this heat is constantly radiated, warming the air with which it comes in contact. The air also receives heat from the sun's rays while they are passing through it. So the atmosphere receives heat in at least three ways: (*a*) directly from the sun; (*b*) by coming in contact with the heated surface of the earth; (*c*) by absorbing the heat of reflected sunlight.

At sun-down, two of these sources of heat are cut off; but, because of the radiation of the heat absorbed by the earth during the day, it does not get cold as soon as the sun is down, and the early evening hours are nearly as warm as just before sun-down. Heat continues to radiate all night, so that it would be colder just before sunrise than at any other time during the twenty-four hours, if all other conditions remained unchanged.

After sunrise the temperature gradually increases. The heat that comes from the sun first evaporates the moisture that has accumulated during the night, and then begins to warm the ground, which, in turn, warms the air; and it is its chief source of warmth, for the sun's rays do not lose more than one-third of their heat while passing through the atmosphere, so the remaining two-thirds must find its way back into the atmosphere by radiation and reflection. This is the reason why the lower strata of the atmosphere are so much warmer than the upper, also why it is frequently so much warmer about 2 o'clock in the afternoon than at any other time during the day, if all other conditions have remained unchanged; for, though the direct heat of the sun decreases after it has passed the meridian, the earth continues to radiate the heat accumulated during the afternoon, which adds its force to that of the sun's rays, this combined heat reaching its maximum about 2 o'clock.

Where there are small hills and valleys, the hills will be warmer than the valleys; for the cold air, being heavier than the warm, finds its way into the lower levels, while the warm air ascends, and adds its heat to that of the hill-tops.

The ocean is not subject to such sudden changes as the land; for the heat received from the sun penetrates it to a greater depth than it does the land, the solar rays affecting water to a depth of 500 feet. Their warmth is distributed through a comparatively thick layer, which it would not be possible to raise to the same temperature with the same amount of heat as it would the thin layer of earth that receives the heat of the sun; it also requires double the amount of energy to warm water that it does to warm the materials of which the land is composed. If the surface of water is cooled in the least, it becomes heavier, and, b cause of the mobility of its particles, immediately settles, and the warmer particles from below take its place, so that ordinarily no portion of a body of water will be warmer than its surface. Since, then, it is impossible for the ocean to acquire its temperature as easily as the land does, it must follow the general law, and retain it longer than the land.

Heat stored up in any substance is called Latent Heat; and, when liberated, it is called Sensible Heat, —that is, the heat that we feel. Water is a great store house for heat; and, when it loses its heat till its temperature is reduced to 32 degrees Fahrenheit, it becomes solid ice. If the water is of the same temperature as your hand, it must give up a number of degrees of heat before it can freeze, and this heat would help to warm all bodies with which it came in contact. In this way ocean currents warm northern latitudes with the latent heat that they have stored up in the tropics.

The ocean is a great modifier of the climate of all of its islands, and of the coasts of all countries bordering upon it; but we must not forget that this surface water may come from the arctic regions as well as from the tropics, and that it may cool the surrounding country instead of warming it. So, in studying climate, we must first determine where the ocean current came from, and the red or blue lines on your globe will tell you; but, whatever its source, it must be warmer than 32 degrees Fahrenheit, and, if the temperature of the surrounding country is lower than that, it will make it warmer. Although the Faroë Islands are in latitude 62 degrees, their inland waters never freeze, owing to the west winds and the sea.

The influence of the ocean currents is always next to that of the winds; but the ocean currents also help to either warm or cool the winds; for instance, cold, deep sea currents rise to the surface just off the coast of northern California, and cause cold day winds at San Francisco.

One other fact must be taken into consideration. Water, both on the sea and on the land, is being evaporated by the sun's rays, and its vapors store up an immense amount of latent heat. These vapors are carried by the winds, and at last condense in the form of rain, and the latent heat is liberated. You can tell what the effect would be, both at the place where the vapors are formed, and where they are condensed.

With all of these different factors at work, it seems as though there never could be the same kind of climate from year to year; but, if the weather for any locality be carefully observed for a long period of time, it will be found that it nearly repeats itself; and, while the years are not exactly alike, some being hot and dry, and others with copious rains, still each country has its own peculiar climate, and it changes but little from year to year. Its extremes of heat and cold may be very great, or its temperature and rainfall may be nearly uniform.

Near the equator, there is no division of the seasons into summer and winter, for the variation of the temperature between day and night is greater than that between the different days of the year; but, no farther than 40 degrees from the equator, this variation may be very great, for in summer the temperature is sometimes 100 degrees in the shade, while in winter the thermometer has been known to fall to 30 degrees below zero, and, as a general rule, the farther from the equator, the lower the winter temperature.

If you will take your globe and place the earth at midwinter's day, you will notice that the most northern rays of sunlight extend only to the polar circle; and, if you will measure the distance from Chicago to the polar circle, you will find it to be about 25 degrees. Then, if you will place the earth at midsummer's day, you will see that the north pole is $23\frac{1}{2}$ degrees from the limit of sunlight, and farther from the direct rays of the sun than Chicago is on midwinter's day. So you can see that one acre of ground at Chicago will receive more light and heat on midwinter's day than an acre at the north pole receives on midsummer's day; and then, taking into consideration the long winter night of the polar regions, and comparing with our winter, which must be warmer than their summer, you can better understand the intense cold of a winter within the arctic circle.

Extensive areas of land must have a lower average temperature than extensive areas of water having the same latitude, and places sheltered from cold winds must be warmer than places exposed to them. Places on different sides of mountain ranges often do not have the same climate. The mountain tops of these ranges may condense all of the moisture of the air, so that the plains on one side, at least, may be deprived of rainfall for a part or the whole of the year. Study the general directions of the winds, and the nearness of the mountains to the sea, and you can determine this feature of climate.

With your globe and what is said above, answer the following questions, and see what your geography says about the climate of the places mentioned. How near can you determine it?

QUESTIONS ON CLIMATE.

If you will look at the analemma on your Tellurian, you will notice that the sun is in the same latitude for a part of February as it is in October. Why is it not as warm in February in the northern hemisphere as it is in October? What part of this chapter tells why?

Cuba and the Sahara Desert have the same latitude. Would they naturally have the same kind of climate? Tell what would naturally influence the climate of each place. (The arrows give the general direction of the wind.)

In the western part of the United States is a large basin surrounded by mountains of considerable elevation. What would be their effect on the rainfall of this basin? What would you expect to find on the mountain tops? How would this affect the winds on the plains below? Would it affect the nature of the soil? The vegetation?

Italy is nearly surrounded by water, and has the Alps on the north. How would this affect her climate? With the desert region at the south of the Mediterranean sea, and with mountain ranges running east and west in central Europe, how would you expect the climate of southern Europe to correspond with that of the United States in the same latitude? What are the climatic conditions of this portion of the United States?

Would you expect the general temperature of South America would be warmer or colder than that of North America? Why? Which would be the more uniform? Why?

Can you give any reason why England should have a warmer temperature than New York? Why it should be colder? Which of these conditions prevails?

Which coast of the United States is the warmest? Why? What portions of the United States have most snow? Why?

The following newspaper item is of interest, and will help you to better understand some of the weather changes that occur in the southern hemisphere while we are having our winter. You will notice that the time referred to was only a few days before our midwinter's day. The year was 1893.

DAYS OF EXTREME HEAT.

Suffering from Torrid Weather in New South Wales.

Great heat prevailed in New South Wales early in December. On December 5 a northwest wind came like a scorching gust from an oven, and early in the forenoon, at Sydney, the thermometer rose to 90 in the shade. By noon it was 93.6, or 54 degrees higher than the record for the year, and it continued at over 93 for fully 2 hours, and was over 90 for three hours. The force of the wind rose from ten to twenty miles, and there was no escaping it.

Again, on December 10th, a wave of heat swept over the colony; the recorded temperatures at many places are said to be higher than ever known. For forty-eight hours it blew from the northward straight down from the tropics across the arid plains of the interior, until the heated air became quite unbearable for white people. At Euston, in the far southwest of the colony, it was 116 degrees in the shade, there being no recognized reliable method known to science for measuring what is known as "in the sun" temperature. The shade heat at Euston would, however, more than satisfy the ardent admirer of hot weather. Balranald reported 111 in the shade; Bourke, 109; Braken Hill, 109, and Deniliquin, 110. Twenty-four stations report from 90 to 100, and temperatures from 89 to 90 were general. At Sydney the weather was extremely oppressive, but the temperature was not within 10 degrees of the heat recorded in the hot northwester of the previous week.

CHAPTER XIV.

THE MOON.

No object in the heavens has attracted so much attention as the *Moon* and her mysteries, for the moon is so near, as compared with other heavenly bodies, that it has seemed as though she must give up her secrets. An object of beauty, "The Queen of Night," she has been the first of the "other worlds" to attract the attention of the child, and call forth its admiration and speculations; and to know her history has been the dream of many an astronomer who would have given his remaining days if he could have shown the world what has transpired on her fair face since first her pilgrimage around the earth began. If the "Man in the Moon" would only tell his story, what audience of interested listeners would equal his?

With the exception of a few days in each month, she stands guard over the sleeping world for a part of each night in the month. For a few days she retires from sight, and hides herself in the glare of the sun's rays, and then comes out with silvery crescent in the western sky to begin her night watch again. And many the glance of love and admiration turned to this new moon.

In Chapter VIII. I told you about the day and night on the moon, and now I want to tell you what must happen during that day and night; but first read again what is said in Chapter VIII. Note what is said about the *phases* of the moon; that is, the new moon, the first quarter, the full moon, and the last quarter, or the old moon, as it is sometimes called. But these phases have nothing to do with what happens on the moon's surface, and that is what we want to consider now.

All of the evidences known to science indicate the absence of any *atmosphere* around the moon, and there are no indications that there is any *water* on her surface. With one of her days equal to about fourteen of ours, with no winds or ocean currents to distribute the heat, and without an atmosphere to absorb one-third of the warmth of the sun's rays, as is the case with the earth, imagination can form but a faint conception of what the heat of a lunar day must be. When we think of the temperature of our hottest summer days, and then think of that temperature augmented by the accumulations of fourteen days of constant glaring sunshine, we have the conditions that are supposed to exist on the surface of the moon during one of her days.

The radiation of heat from the moon's surface can also go on more rapidly than is possible on the earth's surface. In the case of the earth, the air acts as a blanket, and tends to retard the radiation of the heat, while the absence of an atmosphere on the moon's surface permits radiation to go on rapidly, and heat is lost as fast as it has been acquired, and the opposite extreme will be the sure result. The temperature of the night will be as low as that of the day has been high. The intense cold of the long night, about fourteen of our days and nights, can only be compared with the winter season at the poles, unless there are other conditions, unlike those with which we are acquainted, to modify it, and that are quite opposite in their effects to nature's laws on this earth, which is not probable. This seems to settle the question of whether there is any life on the moon or not; but there may be forms of life and intelligence compatible with these conditions. We merely know that, with our present environments, the above conditions would result in the extinction of all of the forms of life with which we are familiar.

We know, however, that life can be sustained under conditions that would be fatal to us. On our earth, life exists everywhere except in the midst of fire, and everywhere Nature has placed some form of life to be supported. "We know, that, in the strong acids that would instantly kill bird, beast, or fish placed within them, there exist and thrive minute creatures, adapted by nature to the strange conditions in which they are placed. Even in the bowels of the earth, and in the very neighborhood of active volcanoes, we find the volcano-fish existing in such countless thousands, that, when they are from time to time vomited forth by the erupting mountain, their bodies are strewn over enormous regions, and, as they putrefy beneath the sun's rays, spread pestilence and disease among the inhabitants of the neighboring districts." So we cannot say that there are no forms of life on the moon, but can say that the conditions on the moon are such that we can only guess at the probable results.

Notwithstanding the probable effect of the sun upon the temperature of the moon during the day, she is spoken of as the "frozen planet," also as the "dead planet." Without vegetation such as we have on our earth, and without life such as we know, the fair face of the moon can be but a dreary waste; still, in the economy of nature, she has probably held an important place and may be to-day as important as the earth, which, compared with the universe, is a mere point in the realm of the unknown.

Our telescopes reveal the fact that there has been volcanic action on the moon the same as there has been on the earth; that her surface is covered with the craters of extinct volcanoes, and that there are hills, valleys, and mountain ranges there, the same as here, which indicates that at some time the moon has been subject to changes like those through which the earth has passed; but the question of what form of life existed there remains unanswered.

The part that the moon takes in the economy of nature is a mystery. Many are the myths that have been told concerning her relations to the earth and to mankind; but men of learning, so far, have found out but few things about the moon of which they feel positive. They know that she is the cause of the eclipses of the sun, and that she is also the principal cause of *Tidal Waves* on large bodies of water.

Tidal waves are caused by the force of gravity in its relation to the moon and the earth. They are also caused by the sun in the same manner as by the moon; but, owing to the comparative nearness of the moon, her influence is greater than that of the sun, and she causes much larger tidal waves than the sun does. The largest tidal waves occur when the sun and moon appear at the same point in the heavens, when their tidal waves unite, producing a wave equal to both that of the sun and that of the moon, and it is called the *Spring Tide*. This tide occurs twice each month, once when the moon is "new," and once when the moon is "full;" but, when the moon is full, the sun and the moon are at opposite points in the heavens, and it is difficult to understand how their tides come together at this time.

The force that causes tides is a constant one; but it does not act equally at all points on the earth's surface at the same time, and but twice in the same place in each twenty-four hours with very nearly the same force. Of course, greater force is exerted on the moon's and on the sun's meridians, and this force is greater when they both are on the same meridian at the same time. The daily rotation of the earth on its axis brings the sun to the same meridian in about twenty-four hours; but the moon does not come to the same meridian in that time, for the moon moves in an easterly direction in her orbit, at the rate of about thirteen degrees for each day; so that, in order to bring the same point on the earth's surface back again to the moon's meridian, it would require a little more than one revolution, enough more to equal the thirteen degrees that the moon has moved in her orbit. It requires fifty-two minutes to do this; hence the tides caused by the moon are fifty-two minutes later each day, while those caused by the sun occur at the same time daily. This would cause their tides to become separated, and they would get farther apart each day for the first week; while, during the second week, the tides caused by the sun would lessen this distance, and at the end of the week would overtake the tides caused by the moon. The moon would now be full, and spring tide would occur again. During the third week the tides would again separate, and at the end of the fourth week they would be together again. At the end of the first and the third week the tides are at their lowest point, and are called *Neap tides*. At this time the sun and the moon partially neutralize each other's tides on the earth, and the tide caused by each is diminished by the effect of the other, as will be shown on page 113. But to understand just *how* these tides are produced is not so easy, and especially how the sun or the moon can produce a tide on the opposite side of the earth as well as on the side that is under their direct influence. The following figure will be of assistance in studying into this matter:

THEORY OF THE TIDES.

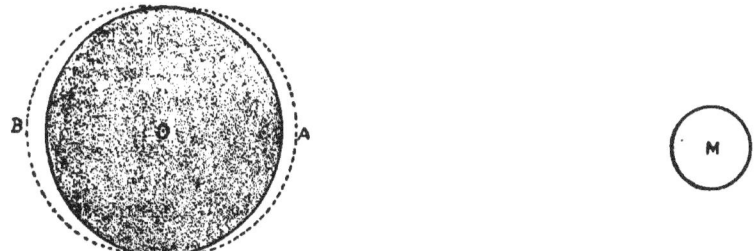

It is not possible to get the proportions accurate in the figure; still, the outlines given will serve the purpose of illustration. M represents the moon, and O the center of the earth. The dark part represents the solid portions of the earth; the light part, within the dotted lines, represents a section of the Atlantic Ocean, and a section of the Pacific Ocean opposite to it, and A and B are opposite points in these sections, and in line with the center of the moon.

No theory for the causes of the tides, yet presented, is entirely satisfactory to the author; but the following is offered as a basis for discussion.

The line of attraction between M and O must be greater than between M and any other portion of the earth's surface; so this would lessen the force of gravity at A, a point on the moon's meridian and on this line of attraction. Gravity being lessened at this point, the tendency would be for the mobile portions of the earth's surface to recede from the center of the earth, producing a long wave in the larger bodies of water. As the sun or moon advances, these waves follow. They are not like waves produced by the wind, however; for they are produced by strong currents, in which water is carried for long distances on the crest of the wave, and backward long distances in its trough. That tidal waves are caused by a lessening of the force of gravity, is in harmony with the fact that water is affected throughout its entire depth, while other waves disturb water to only a small depth below the surface.

The tidal wave at A, and the moon's attraction, would naturally move the center of gravity, O, nearer to A than the center of the earth is, and that in turn would lessen gravity at B, a point opposite from A.* The result would be a tidal wave at B, which would maintain an equilibrium of the earth in her daily rotation. These tidal waves are like the day and night hemisphere in their course around the earth, for it is the earth's rotation that carries them around her.

When the moon is in quadrature the tides are at their lowest, for then the tides of the moon occur in the trough of the tides produced by the sun. The following figure will make clear what is meant by

* Read the theory based on the revolution of the moon and the earth around their common center of gravity.

the crest and trough, the ebb and flood, of the tides:

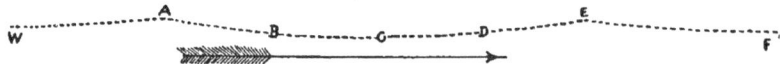

The arrow points in the direction of the rotation of the earth, and the direction of the tides would be just the opposite; that is, from east to west. The points A and E are at the *crests* of the tidal waves, and C is at their *trough;* D is at the *ebb*, or falling, of the tide, while B is at the *flood*, or the rising, of the tide. It is said to be *high tide* at A and at E. and low tide at C.

Refer again to what is said on pages 109 and 110, keeping the above figure in mind. First consider spring tide at A, and then consider that it is a week later and that neap tide is at A. Under these conditions where would the sun's tide be?

While the tides produced by the sun occur about twelve hours apart, those produced by the moon occur about fifty-two minutes later each day; so that, if the crest of the tide produced by both the sun and the moon, were at A, in a week's time the tide produced by the moon would have fallen behind to the point C, and would then be in the trough of the sun's tides. It would continue to fall behind again till it reached the point E, when it would be at the crest of the sun's tide again, and would produce a *Spring Tide.* When at C it is called *Neap Tide.*

The shape of the coast lines of the oceans has a marked effect on the height of the tides, while in mid-ocean it is probably less than two feet and hardly noticeable. At the Bay of Fundy the tides rise to their greatest height on the globe, the Spring tide rising to a height of 50 feet, and the Neap tide 24 feet. In mid-ocean the highest point of the tides would follow a line drawn from the center of the moon to the center of the earth; but, when the crest of the tide reaches a coast line, the direction of the tide is changed, and takes the direction of the coast line. If you will take your globe and find the Bay of Fundy, you will notice that it is a long, V-shaped body of water, and that it opens into the ocean in such a way that it will catch the tide as it turns off from the coast of the New England States and goes north. The water becomes hemmed in as it approaches the point of the V, and the effect is the same as it would be if the current of a large stream of water should be stopped by a dam.

When the tidal wave crosses the ocean from the east, and reaches the western coast, it is deflected from it, and takes the direction that offers the least resistance to its onward flow, as I have stated above.

When it can turn in neither direction, it piles up on the coast line till the tidal influence has passed, and then flows backward, taking the name of *Ebb Tide*.

We often hear of the "man in the moon," light and dark patches on the moon's disc, which are probably caused by irregularities of the surface, and have caused a great many to imagine that the surface of the moon resembles a man's face. When the moon is in quadrature, she presents a beautiful appearance, and, when viewing it with a good telescope, these irregularities are quite distinct.

Handed down to us with other myths that have found their way into every language, is the belief that the moon in some way affects the weather and the growing crops. It is not uncommon for some farmers to wait for a certain phase of the moon before planting seed, firmly believing that it is an important step in assuring an abundant harvest. They watch the new moon to see whether it is going to be wet or dry, without thinking, that, at nearly the same date of the previous year, the horns of the moon pointed in exactly the same direction as at the time when they are looking to see what the weather will be for the coming month. For them the new moon determines this all-important question.

Before attempting to explain why the horns

of the moon do not always point in the same direction when the moon is new, I want to call your attention to the figure on the following page, showing the *nodes* of the moon, and the inclination of the moon's orbit to that of the earth.

No attempt has been made, in this figure, to get the exact proportion as regards distance. That would be impossible and have the size of the moon such as would show its crescent to advantage, and the figure as it is, will serve all purposes. The actual inclination of the moon's path to that of the sun is five degrees, while the angle at which the lines in the figure cut each other, is fifteen degrees; but the distances are so great, and the sun and moon so small in comparison with surrounding space, that the phenomena shown in the figure occur as represented. The arrow at the right margin represents direction with reference to the earth's surface, and the dotted lines represent the paths of the earth and the moon as indicated. The star is at the moon's *node*, that is, the point where her path crosses the sun's path. Since the moon's revolution around the earth occurs once in twenty-nine days, she must cross the sun's apparent path twice during this time, and half of her journey is north, and half of it is south, of the sun's apparent path. The points

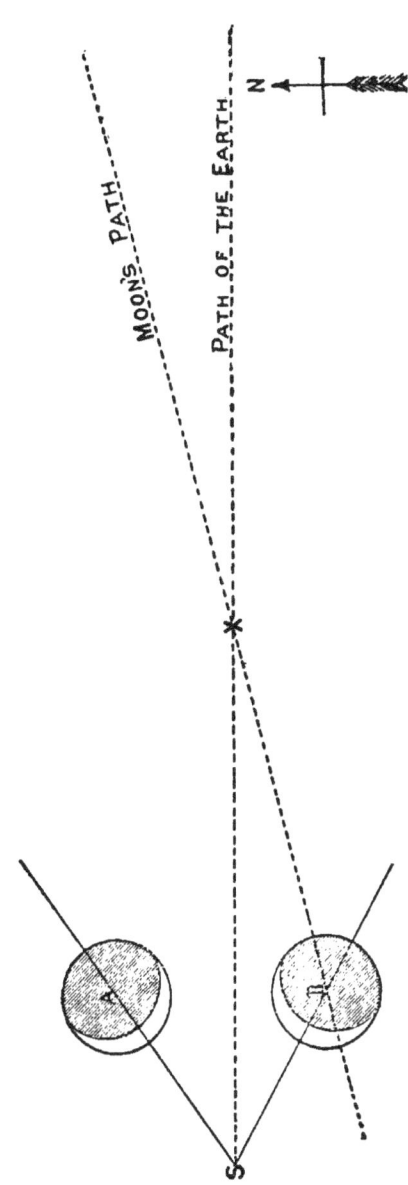

NODES OF THE MOON.
(117)

where the moon's path cuts the ecliptic (the sun's apparent path) are called *Nodes*. The node where the moon passes from the south to the north side of the ecliptic, is called the *Ascending Node*, and the opposite node is the *Descending Node*.

If the new moon occurs while she is in her ascending node, the horns will be in the direction that they are when the moon is at A; but, if the new moon occurs when she has passed her descending node, the horns will point in the direction that they do in B; the horns always pointing in directly the opposite direction to that of the sun from the moon. So, by looking at the moon, you can tell the direction of the sun from her, no matter at what time of the night you look. When she is south of the sun's path her horns will point downward; and, when she is north of the sun's path, they will point upward. These phases of the moon are called the *wet* and the *dry* moon. I think that the moon in the ascending node is called the *dry* moon; for I can remember, when a little boy, of hearing my father say, "The old Indian could hang his powder horn on the moon to-night, and it wouldn't slip off, and I guess that we won't have much rain this month;" and my good old grandmother used to say, "that the horns of the moon point upward, and that will hold all the water in the hollow, so the earth will go dry this month."

S marks the position of the sun, and the lines drawn through A and B show the direction of rays of light falling upon the moon. Of course, the side of the moon that is toward the sun is the part that will shine, and we can see only a small crescent-shaped strip of that part. A line drawn through the points of that strip will pass through the center of the moon, and be at right angles to the rays of light that come from the sun. When looking at the moon, draw imaginary lines as I have them in the figure, and it will help you to locate the sun, and tell you whether the moon is in her ascending or descending node; that is, whether she is north or south of the sun's apparent path. The latitude of the observer will make a large difference in the direction of the points of the crescent. If you will look at the moon, some time, at about 3 o'clock in the morning, you will be surprised at the direction in which she indicates the sun to be. There is another interesting thing about the moon that I want to call your attention to before closing this chapter.

You have probably noticed, that, at some times of the year, when you are on the moon's meridian, she appears to be much nearer to the southern horizon than at others, and that in the winter season the moon is almost over our heads when we are on her meridian. I can remember the old weather prophet saying: "The moon runs high to-night; look out for cold." Well, that is true too, for the moon never "runs high" except in midwinter, as you can see by looking at your globe, and a clear midwinter's night is pretty sure to be cold in this latitude.

Place your globe at midwinter's day, and fix the moon so that it will be directly opposite to the sun. Rotate the globe so that Chicago will be on the sun's meridian, and notice the angle that a line drawn from Chicago to the center of the sun will make. Then rotate so that Chicago will be on the moon's meridian, and draw a line from Chicago to the center of the moon, and notice that this angle is not so great as the other; that is, that the moon is more nearly overhead, and "runs high." Place your globe on midsummer's day, and try the same experiment. Remember that the moon can never get more than five degrees farther north than the sun, and that she cannot go more than five degrees farther south than the sun's path; so she will appear sometimes nearer the southern horizon than the sun ever does; and, when the sun is nearest to the southern horizon, the moon is the nearest to the zenith; that is, when the moon is full. Keeping in mind that the paths of the sun and moon are never more than five degrees apart, this apparent difference will be, to you, one of the best proofs of the inclination of the earth's axis to her orbit; for you are in the same relative position to the moon on midwinter's day that you are to the sun on midsummer's day, if you but make allowance for the angle of five degrees that denotes the moon's inclination.

Observation will make you familiar with all that I have said about the phases of the moon, the "wet" and the "dry" moon, the "high" and the "low" moon, and the moon's nodes. You will not need instruments for this work. Observation of the moon will also assist you in locating most of the planets in our solar system, and will help you in recognizing them by name. Get an Ayer's Almanac, and study carefully the characters that represent the different planets and their positions in relation to the earth, the sun, and the moon. I have before me their 1894 almanac (see "Appendix"), and make the following observations regarding the month of January. On the 3d of the month I find that Mars is in conjunction with the moon; so, by looking at the moon on that date, I shall see Mars near her, and two days later Mercury will be the nearest planet to the moon. Now, in two days, the moon has traveled twenty-six degrees in her orbit, hence Mercury and Mars are about 26 degrees apart at this date; and, by watching them from time to time, I will see that this relative position will change. The moon is also at her most distant point from the earth at this date.

On the 10th, the moon is in conjunction with Venus; and at this time Venus is at her nearest point to the earth, which would cause her brilliancy to be at its greatest. On a clear night the star and the crescent in the western sky are a beautiful sight.

On the 13th, the moon begins her ascending node, and on the 16th is in conjunction with Jupiter, one of the most interesting planets in the heavens. ' On the 17th she is in conjunction with Neptune; so Jupiter and Neptune are about 13 degrees apart on this date. On the 19th the moon has reached the most northern point in her ascending node, and is the nearest to the earth that she gets. She does not overtake any more of the planets until the 27th, when she is in conjunction with Saturn, and on the 31st has completed her orbit and is in conjunction with Mars again. So, in making her circuit of the heavens, she has passed all of the planets, and you can get their names by watching for her conjunction with them.

You probably noticed that she began her descending node on the 26th, and, of course, her different phases have not escaped your attention. You will notice that in March and August the moon is nearer to her ascending node when new, and to her descending node when full, than at any other time during the year. These are the conditions necessary for eclipses, which we will study in another chapter. Notice how the time of the new moon and the ascending node vary from month to month. The same is true of the full moon and the ascending node.

You will see, that, by a little observation, you can become quite well acquainted with the heavens, and it will be with pleasure and interest that you will watch the ever-changing positions of the moon and the planets. You will feel that you are in closer communion with the mysteries that surround you, and in some way you cannot help feeling that you are a part of some plan that is being worked out by a Power that you seem conscious of, but can but very imperfectly know under present conditions. We can only stand between the two eternities, and wonder at the immensity of the problem before us, and feel, that, in some way, we must take part in its solution.

Let us see what we can gather from the sun.

CHAPTER XV.

THE SUN.

The Sun is the center of the solar system, around which the earth and all of the other planets revolve. He has an atmosphere, and is composed of the same kind of materials as the earth and all of the other planets that revolve around him; in fact, the earth and all of the other planets are supposed to have once been a part of the sun, and to have been evolved from his surface from time to time, in the order of their distance from him. One cannot imagine the immensity of the sun before any of the planets were detached; for his size at the present time is beyond our comprehension, his diameter being estimated at 852,584 miles,* while that of the earth is a little less than 8,000 miles, his volume exceeds that of the earth by 1,245,126 times, and there is enough matter in the sun to make over twelve hundred thousand worlds like ours.

His distance from the earth is 91,430,220 miles,

---

* These figures are given in round numbers. Different texts do not agree on the exact numbers.

and, to the naked eye, he appears only as a luminous mass of intense and uniform brightness, giving both light and heat to the solar system.

Before the evolution of any of the planets, the entire substance of the sun is supposed to have occupied all the space included in the orbit of his most distant planet, and to have appeared in the heavens as a nebulous cloud, perhaps as the nebula in Orion appears when viewed with a telescope. This nebula is one of the most beautiful objects in the heavens.

There are many things about the sun that afford subjects for speculation, and one of them is the spots that appear on his surface from time to time. By carefully observing them for a long period of time, it has been determined that the sun rotates on his axis once in twenty-five of our days. The cause of the spots on the sun has not been determined, and a satisfactory description of them cannot be given in a book like this. They vary from time to time, both in number and size; but there is seldom a time when, by the aid of a telescope, they are not to be seen, and in some years they can be seen every day. In 1860, there was not a day when there were no spots on the sun's disc, and during the year 211 different groups appeared, while in 1867 there were

195 days that the sun's face was without a blemish.

The spectroscope has told wonderful stories concerning the sun; and you will be well paid to find out what this wonderful instrument is, and how it has been able to tell so much about the sun and the stars. What it has told verifies what had already been supposed to be true; that is, that the earth is composed of the same kind of materials as the sun, and that the sun is so intensely hot that it is capable of supporting iron, copper, and some of the other metals in the solar atmosphere, just as water is supported in the atmosphere of the earth. At some time the sun will get cool enough so that these metals will be precipitated the same as rain is precipitated on the earth. Geology teaches that the earth has passed through just such a stage of development; that at some time the materials that now compose the earth's crust were held by heat in suspension just as the clouds are; and that, when sufficient heat had been radiated into space, these materials descended just as the rain does now, and that the changes that have gone on since have been more or less dependent on the radiation of heat into space. What ultimately becomes of the heat is an interesting question, and one that will afford ample room for study.

The spectroscope also tells us that the sun is composed of the same kind of materials as the stars; and our earth being made from these materials also tells us that we live on one of many worlds that form the universe, and that we, as a part, bear some mysterious relation to the whole. But, so far as we can tell, the same conditions do not exist on any of the other worlds that we have on ours; and, if the theory of the formation of the earth is correct, we know that there are many worlds that have conditions far different from ours; and, while we cannot help thinking that they support some form of life, we cannot help wondering what that form of life may be. That so many worlds should be created without purpose, or that they came by chance, is not to be thought of for a moment. That all should be made to serve the purpose of one small world, our earth, seems almost absurd; for it is impossible for us to comprehend how even the nearest planet of our system can be necessary to our existence, or how it can in any way serve us, except as a thing of beauty,—one of the gems of the night. We cannot even tell the object of the creation of our own world.

The stars, being in the same condition as the sun, cannot support forms of life such as we know; but, like the sun, they may be centers of systems of planets like ours, peopled with beings like ourselves.

This seems more than probable; for, being alike in so many respects, they must, in some way, contribute to the evolution of a higher form of consciousness than that which we know, and to which we hope to attain.

When we think that the sun is radiating into space, in all directions, light and heat, and then think what a small amount, compared with the whole, reaches any of the planets, we cannot help but wonder at the seeming waste of energy, and the inquiry naturally arises as to what becomes of it. Is it distributed to every point of the universe? Will the sun continue to lose his heat until, at some time in the eternity to come, he will become like our moon,—frozen,—and be the center of dead and silent worlds, which, wearied by their ceaseless flight of countless ages, stumble from the paths they have so long followed, and rush together, returning to Chaos, whence they came? Will they form new worlds again?

The subject of light and heat is one full of interest, and one concerning which there has been much speculation. What is light? What is heat? We might say that light is that peculiar form of energy that is recognized only by the visual organs; but that gives no idea of what light really is. The undulatory theory of light is the accepted explanation of this phenomenon; that is, it is a mode of motion produced by luminous bodies; but is the question answered?

I have said that the sun is the source of all light; but I am not so sure of that after all, for one of the most effective forms of artificial light with which we are acquainted, cannot be traced directly to sunshine in any form. I refer to the light that is produced directly by electricity, and electricity has so far refused to give up the secret of its origin. We know how to make its presence known, and can force it to serve us in many ways, but what is it? The undulatory theory of light must include light produced by electricity as well as direct sunlight.

Light, according to the undulatory theory, consists of a wave-like motion that is recognized by the visual organs only, and is produced by the bodies that are called luminous. Through the sensation of light we become aware of the existence of bodies with which we are not in actual contact, and the sensation that they produce is called color. Different colors are said to be produced by different wave-lengths; and the wave-lengths that can produce the sensation of vision are bounded on one side by the red, and on the other by the violet, rays of light. The wave-length of the extreme red is said to be twice that of the extreme violet; and between the red and violet are the wavelengths that produce orange, yellow, green, blue, and indigo, which, with the first two mentioned, are called the colors of the spectrum. When mingled in the proportions that nature gives them, they form white light, by which is meant colorless light.

Outside of this spectrum, wave-lengths are known to exist, for science has been able to locate, by means of photography, stars that the telescope could not reveal; because, while they did not have the power to produce any sensation on the visual organs, still they produced the chemical effect necessary in photography. The existence of rays having wave-lengths twenty times that of red rays has been found in the radiations of the moon.

The difference between light and heat is a difference of wave-lengths only; and both, probably, are forms of the same energy of which electricity is a near relative. If all of these various wave-lengths could produce the sensation of vision, what wonders might be revealed to us!

But it is useless to follow this line further; for, while this theory has borne much good fruit, and while it answers more of the questions concerning light and heat than any other that has so far been offered, it is probable that science will gather a better knowledge of this subject than we have at present, and that new fields of thought will be opened.

In closing this chapter, we can feel assured that the place our sun occupies in the universe is that of a star, that in all probability the stars are centers of solar systems like our own, and that their destinies are subject to the same course of natural events. Nature has not yet seen fit to give up her secret, and she will hold her own counsel till that point in the evolution of consciousness has been reached, where she can share the wealth she has in store with those who thirst for the truth, and have been willing to delve deep for the hidden treasures that are to be found in the storehouse of knowledge.

CHAPTER XVI.

ECLIPSES.

It sometimes happens that the earth, the sun, and the moon get into such a position that their centers are in direct line; and, when this occurs, there is an eclipse either of the sun or of the moon. This occurs twice each year; that is, they get near enough in line to produce partial eclipses. But, before going farther into the subject, I want to give you some diagrams showing the relative sizes of the earth, sun, and moon, so that you can better understand the length of the shadow that is made by the earth and the moon.

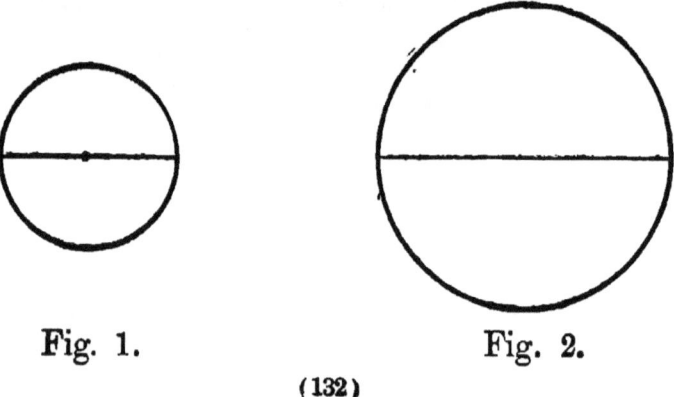

Fig. 1.   Fig. 2.

Fig. 1 represents the moon's orbit, and Fig. 2 the size of the sun; while the small black spot in the center of the moon's orbit, is larger than would correctly represent the size of the earth if drawn on the same scale as the sun and the moon's orbit. It would take sixty worlds like the earth, touching each other, to span the moon's orbit.; yet, as you can see, this orbit is much smaller than the sun. The following figures represent the earth and the moon, drawn to the same scale:

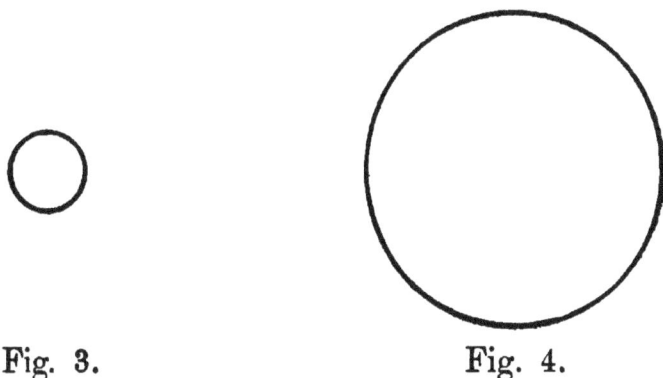

Fig. 3.   Fig. 4.

Fig. 3 represents the size of the moon, and Fig. 4 the size of the earth, drawn to the same scale. Fig. 4 and Fig. 2 are of about the same size, but they are not drawn to the same scale. If they were, Fig. 4 would not be so large as the small black dot in Fig. 1 and the moon (Fig. 3) would be hardly large enough to be seen. It does not seem possible that there is such a difference in the sizes of the earth, the sun, and the moon; but such are the facts, as shown by the number of miles that measure their various diameters, and the preceding diagrams may give a better impression of these facts than the numbers would.

It will be impossible to make a diagram that will give the relative sizes, and at the same time give the relative distances, of the earth and the sun; but we give the following illustration, representing the moon's orbit, and the relative length of the earth's shadow, also the direction of the sun's rays. A careful study of this illustration will tell you why eclipses do not occur each time that the sun and the moon are in conjunction or in opposition.

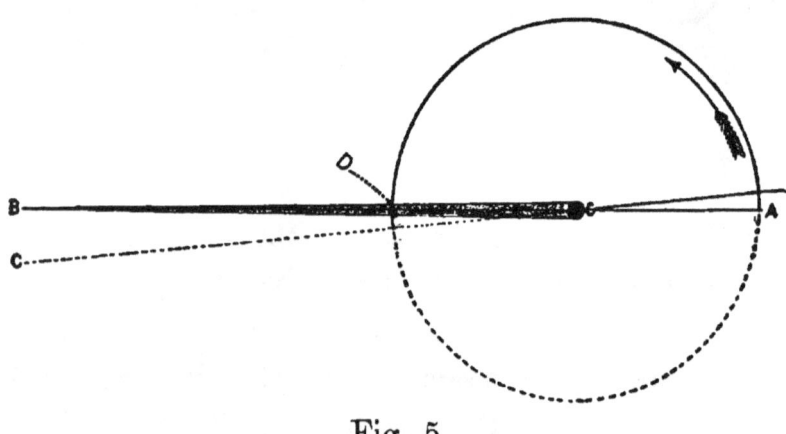

Fig. 5.

The large circle represents the moon's orbit, and the arrow indicates the direction that the moon takes

THE NODES.

in her revolution around the earth, also the direction that the earth rotates on her axis. The points A and D are the moon's nodes. The point A is called the ascending node, and the point D the descending node. The line beginning at C and running through the earth, O, is in the same plane as the moon's orbit. It is drawn at an angle of five degrees to the plane of the ecliptic, and the moon's orbit intersects the plane of the ecliptic at the points A and D. The earth's shadow, which terminates at B, is always in the plane of the ecliptic, which is at an angle of five degrees to the moon's orbit. Suppose that the page upon which this is printed represented the plane of the ecliptic; then the moon's orbit would be inclined to the page at an angle of five degrees, and would pass through the paper at the points A and D. The point C would be on the other side of the paper, and the line beginning at the point would pass through the paper at the point O, and would continue on this side of the paper, passing through a point on the moon's orbit five degrees from the point A.

The little dot at D is really larger than the moon would be, drawn to the same scale that the earth and the orbit of the moon are; but it represents the moon in this figure. You can see that she is just entering her descending node. When she reaches her ascending node, her shadow, which begins at A, will extend toward the earth; but it may not be long enough to reach as far as the earth is from the moon; for, when the moon is at her most distant point from the earth, her shadow is not long enough to pass half across her orbit. Sometimes the moon is 251,947 miles from the earth, her nearest point to the earth is 225,719 miles, and her shadow varies in length from 221,148 miles to 252,638 miles.

Now, in the revolution of the earth around the sun, the shadow of the earth is always in a line which, if extended, would pass through the sun; and this shadow lies entirely outside of the plane of the moon's orbit, except twice each year, but it is always in the plane of the ecliptic. If you will imagine the point A moving around in the moon's orbit, you will see the general direction that her shadow will take; and you will see that an eclipse cannot happen until the earth, in her revolution around the sun, reaches a point that shall be very near to the plane of the moon's orbit. Then, when the sun and the moon are in conjunction or opposition, an eclipse is liable to occur; for then the centers of the earth, the sun, and the moon, are nearly in line. If they are in conjunction, the eclipse will be that of the sun, and its character will be determined by their nearness to the nodes and the moon's distance from the earth. If they are in opposition, the eclipse will be that of the moon, and its character will also depend upon the nearness of the sun and moon to the moon's nodes.

We will first study the eclipses of the sun; and, of course, the sun and the moon must be in conjunction. We will also consider that they are very near the node, and that the moon's shadow will fall on the earth. If she is nearing her ascending node, her shadow will sweep across the south polar region; but, if she is nearing her descending node, her shadow will sweep across the north polar region. The nearer the node the conjunction occurs, the nearer the equatorial regions the field of the eclipse will be. If the observer is in this region, he will see an eclipse of the sun; and, if he is so fortunate as to be in a direct line drawn through the centers of the moon and of the sun, and the moon is at her nearest point in her orbit to the earth, the moon will appear to be as large as the sun, and there will be a total eclipse; that is, no portion of the sun will be in sight. But, if the moon is at the most distant point in her orbit from the earth, and the observer is at the point mentioned above, a narrow rim of the sun will appear to encircle the moon, and it will then be called an annular eclipse of the sun. If the observer is not in the direct line of their centers, but is in the field of the eclipse, the moon will pass over only a part of the sun's disc, and there will be a partial eclipse of the sun. A total eclipse of the sun is visible on only a small portion of the surface of the earth, while a partial eclipse is visible on a larger portion, as can readily be seen by an examination of the following figure, which gives the relative sizes of the earth and the moon, and will also show how eclipses occur:

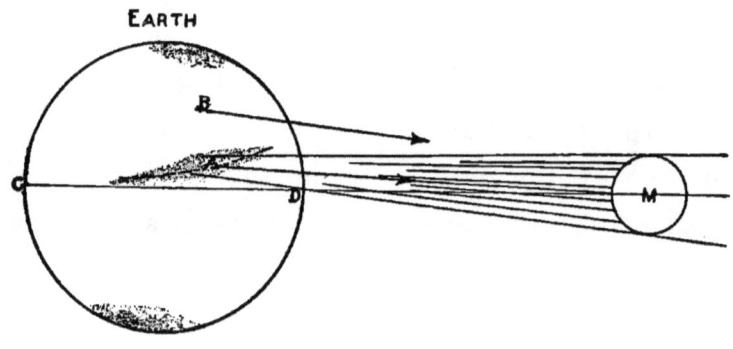

Fig. 6.

If the sun and the moon followed the same path in the heavens, there would be an eclipse every month; but the moon's path in the heavens is inclined about five degrees to that of the sun, and their distance from the earth is so great that they must be near the point where their paths cross each other in order to produce an eclipse. Their paths cross at opposite points of the heavens, called the moon's nodes. The moon passes these nodes twice each month,—once when it is new, and once when it is full; but only twice during the year is the sun near enough to these nodes when the moon passes them to produce an eclipse of either the sun or the moon.

In the preceding figure, the shaded spot on the earth represents a region where a total eclipse of the sun, indicated in the figure, is visible. The observer at A is looking in the direction of the arrow, from that point toward the sun, and the moon is directly in the path of his vision, so that the sun is not visible, but is hidden behind the moon, and the observer is really looking at the night hemisphere of the moon. The sun is at the moon's node; but she advances so rapidly in her orbit that it is rarely more than five or six minutes that the sun remains entirely hidden, and sometimes a total eclipse lasts only a few seconds, and is visible to only a small strip of country; but the time from the first apparent contact of the moon with the sun until the sun's disc is entirely uncovered is from two to three hours, and the region of country in which the partial eclipse is visible is quite extensive. The arrow B indicates a locality in such a region, and at this place the sun would appear something like this:

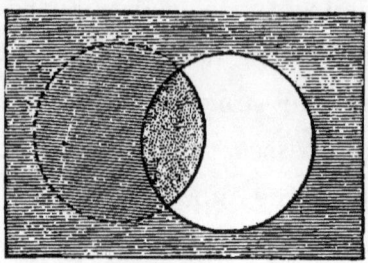

Fig. 7.

An eclipse of the moon occurs when the sun is near one of the moon's nodes, and the moon is at the other. This would bring the moon in the shadow of the earth, which would darken a part or the whole of her disc. Eclipses of the moon can occur only when she is full, and the kind of eclipse depends upon whether she is at the nearest point in her orbit to the earth, or whether she is at one node at the same time that the sun is at the other.

Fig. 7 will illustrate the appearance of the sun during a partial eclipse. A total eclipse cannot be represented by a picture that will give a very good idea of how it really looks, and, for that matter, no picture of an eclipse is a good representation of how it really appears. Fig. 8 will give you an idea of an annular eclipse.

The ring shown in the figure cannot be perfect

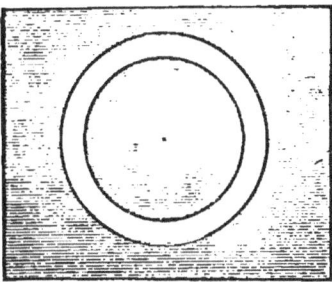

Fig. 8.

more than a few minutes, for the moon moves across the sun's disc in an hour or two. In order to make a perfect ring, the center of the moon must pass along the sun's diameter. The ring will be perfect when a line drawn from the point of observation will pass through the center of the moon and the center of the sun.

We will now see what is liable to happen when the sun and the moon are in opposition, which will occur about two weeks after an eclipse of the sun. I should have stated that an eclipse of the sun can occur only when the moon, at the time of her conjunction with the sun, is within $19\frac{2}{3}$ degrees of her node, and will certainly occur when she is within $13\frac{1}{2}$ degrees of her node. An eclipse of the moon can only happen when she is within $13\frac{1}{3}$ degrees of her node at the time of her opposition with the sun, and must occur if she is within 7 degrees of her node at that time. If you could be on the surface of the moon at the time of her eclipse and should look at the sun, the earth would seem to get between you and the sun, and there would be an eclipse of the sun. It would look just the same as it does when viewed from the earth; but an observer on the earth's surface, watching the same point on the moon at the same time, would see the moon entering the earth's shadow, which extends more than three times farther from the earth than the distance to the moon. Of course you will remember that the moon is full, and that her diameter is 2,160 miles, and that she may not be near enough to her node to permit more than four or five hundred miles of her diameter to pass through the earth's shadow. She would then appear like Fig. 7. If the moon should happen to be exactly at her node at the time of the full moon, a total eclipse would occur; that is, the entire surface of the moon would enter the earth's shadow; but the moon moves so fast in her orbit that its duration would be for a short time only.

This year, 1894, there are four eclipses,—two of the moon, and two of the sun. The first is a partial eclipse of the moon, the 21st of March. It is visible, more or less, to the extreme western portion of North America, Asia, Australia and the Pacific Ocean.

A PARTIAL ECLIPSE.

About two weeks later, the 6th of April, there is an eclipse of the sun, visible only to eastern Europe and Asia. It is an annular eclipse.

The next season of eclipses is in September, and on the 14th and 15th there is another partial eclipse of the moon, visible to North and South America, the western portions of Europe and Africa, and the Atlantic and eastern portion of the Pacific Ocean. Two weeks later there is a total eclipse of the sun visible only to portions of Africa, Hindostan, Australia and to the Indian Ocean.

If you thoroughly understand all that has been said in this chapter, you can answer the following questions. Some of them can be answered by carefully remembering what has been said in the text; but, to answer all, you will be obliged to use your imagination, and you can get some valuable assistance by the aid of your Tellurian.

QUESTIONS.

1. Is the moon in her ascending or descending node at the time of the first eclipse?

2. If she had been exactly at her node, what would have happened? Is she at her farthest limit from her node, or near the node?

3. Can you tell about what time of day it is by the location of the territory in which the eclipse is visible? Can you tell, by the extent of this territory, about how long it took the moon to pass the earth's shadow?

4. What would have been the necessary conditions for this eclipse to have been visible in the entire United States? For it to have been seen in South America?

5. In the eclipse of the sun, is the moon at her nearest or most distant point in her orbit from the earth? What would have made this a total eclipse?

6. In what node is the moon at this time? Can you tell the time of day that this eclipse occurs?

7. At what time of day does the second eclipse of the moon occur? Will the shadow first appear on the northern or the southern limb of the moon? (Limb means the border or edge of the disc of a heavenly body.)

8. Can you tell anything about what time it is in the United States when the last eclipse of the sun occurs? Which of the above questions will apply to this eclipse?

With an 1894 almanac you can tell how nearly correctly you have answered these questions. You will probably need one to answer the last question, but you should be able to answer many of them with the aid of the Tellurian.

In the next chapter we shall take up the discussion of time.

CHAPTER XVII.

THE CALENDAR.

We have seen that the rotation of the earth on its axis, causing day and night, makes a natural division of time. The revolution of the earth around the sun makes another natural division of time that has served the purpose of fixing the dates of events for ages. It seems probable that the need of some exact method of reckoning time first gave rise to the study of astronomy, and called attention to the movements of the planets.

The revolution of the moon around the earth has also served to help keep a record of time, each lunation having been called a *month*, until the establishment of the Julian Calendar, and twelve lunar months still make a year with the Mohammedans. But the revolution of the moon around the earth is not an exact measure of the revolution of the earth around the sun; so the year has been made the basis of reckoning time; and, in order to make the twelve months an exact measure of the year, various changes have been made in the number of days that should make the different months. Instead of being of the same length as the old lunar month of thirty days, a day has been added to a part of them, and two days taken from one of them, except in leap year, when only one day is taken from this old lunar month; or, as we say now in leap years, we add one day to February.

Our calendar, which is called the *Gregorian Calendar* because of certain changes that are made according to the directions of Pope Gregory XIII., is in general use, except in Russia, where the old Julian calendar is still used. It is but little changed from the Julian, the most important change being that of dropping the leap year from the centuries that are not exactly divisible by four; for instance, 1700, 1800, and 1900. The average calendar year is still a few seconds longer than the tropical year; but this will not make a day's difference till over 4,000 years have passed.

The tropical year is 365 days 5 hours 48 minutes and 49.7 seconds. You will notice, with the aid of your Tellurian, that, in each year, the earth makes one more revolution on its axis than the number of days in a year, making a little more than one revolution each day on account of its onward motion in its orbit around the sun. You will notice, too, that, owing to the distance traveled by the moon in her orbit each day, it takes more than twenty-four hours for the same meridian on the earth's surface to return to the moon's meridian.

The number of days in the week does not seem to have an exact relation to any of the other divisions of time; but the weeks seem to be, in their nature, religious divisions, and were probably established to meet the religious requirements of man.

The *Dominical Letter*, the letter of the alphabet that denotes Sunday, probably has a history that dates back to some religious rites for its origin. The days of the week and other divisions of time, and the dominical letter seem to bear a strange relation to each other, that, without their history, is not easy to understand. The following couplet, with the methods of calculation described in the rule and problems, will give the day of the week upon which any event happened if the date is known:

Jan.	Feb.	March	April	May	June
A	D	D	G	B	E
July	Aug.	Sept.	Oct.	Nov.	Dec.
G	C	F	A	D	F

A	B	C	D	E	F	G
1	2	3	4	5	6	7

THE CALENDAR.

The following mnemonic will assist you in remembering the couplet by using the first letter of each word:

> At Dover Dwells George Brown Esquire,
> Good Carlos Finch, and David Friar.

G is the dominical letter for 1894, as found by following the directions given in the rule below:

RULE.

I. To the number of years above centuries add their one-fourth (omitting fraction), plus 2.*

II. Divide this sum by 7. If there be a remainder, subtract it from 7, and the result will be the number of the dominical letter. If there be no remainder, the number of the dominical letter will be 7.

III. In the foregoing couplet, the letters above the line represent the months. A is January, D is February, D is March, G is April, etc. The letters below the line represent the days of the week, the dominical letter being Sunday. The figures are the numbers obtained in the solution of problems for finding days of the week from dates.

IV. Begin with the dominical letter, calling it Sunday, the next to the right Monday, and so on till you come to the letter corresponding to the letter that represents the month given in the "date" required, and the day of the week represented by this letter

* For January and February of leap years add 1.

will be the first day of the month. (See correction for leap years, given with one of the following problems.

PROBLEMS.

1. Find on what day of the week May 28th, 1894, occurs; also on what day of the week were January 16th, 1892, and March 5th, 1892.

SOLUTION.

(a) Years above centuries,   94
(b) Their ¼, the remainder not being considered,   23
(c) Plus   2
                                      ———
                                      119

The sum 119, divided by 7, equals 17 with no remainder. Then the number of the dominical letter is 7, or G.

B, in the couplet, represents May, and is numbered 2 in the list of letters representing the days of the week. Then G is Sunday; 1, or A, is Monday; and 2, or B, is Tuesday. So, the first day of May is Tuesday. The 1st, the 8th, the 15th, the 22d, and the 29th are also Tuesday, and the 28th is Monday.

SOLUTION.

92
23
 1   7 less 4 is 3, and 3 is C, the dominical
—   letter for January and February,
7)116   1892. This would make January
—   16th come on Saturday.
16 and 4 rem.

As 1892 was a leap year, we add to the years above centuries their one-fourth, plus 1. After February we add 2, in determining the dominical letter for the balance of the year.

2. Find the day of the week that Columbus landed in the new world. On what day of the week was the Declaration of Independence signed? The World's Fair closed October 31, 1893. What day was it?

3. On what day were you born? On what day will the first of March, 1900, occur?

Here is another method of finding the day of the week that will furnish you with a fine problem for explanation.

To tell the day of the week, first take the number represented by the last two figures of the number of the year, and add a quarter to this, disregarding the remaining fraction. Then add the number of the day of the month to the sum, and to this add the number standing for the month, taken from the series 366,240,251,361, counting the first figure in the series as January, the second as February, the third as March, and so on. Divide the sum by 7, and the remainder will give the number of the day of the week; and, when there is no remainder, the day will be Saturday.

As an example, take March 19, 1890. Take 90, add 22, add 19, add 6. This gives 137, which divided by 7, leaves a remainder of 4,* which is the number of the day of the week, or Wednesday.

You remember that it is always noon on the sun's meridian; and you have probably taken a trip around the world, in your mind, keeping up with the sun, and found "everybody eating dinner as you passed." Perhaps you wondered how it happened, that it was Monday at noon all of the way around, but that, when you got to the starting point, it was Tuesday noon. This brings up the question of where the old day

---

* The day before the one indicated by the "remainder" will be the day sought in January or February of a leap year.

ends and the new begins. Sailors change their dates at the 180th meridian. If they are sailing westward they add one day to their reckoning, but deduct a day if they are going east. This happens in the Pacific Ocean.

Settlements have been established on the islands of the Pacific by all nations of the world; and they have carried with them the day of the week as it occurred at home, making no change for the change of longitude, so that people coming from eastern countries have established days different from those of people who have come from western countries. It is said that the days in these islands have not yet been made uniform, so that it is Monday at one place, and Sunday at another, on the same meridian.

In *American Notes and Queries*, September 1, 1888, page 215, is the following interesting item, under the heading of "The International Date Line:"

"The international date line is a line at which dates must be made later by one day when crossing it from east to west, and earlier by one day when crossing it from west to east. This line passes just west of Behring Strait, west of St. Lawrence Island, west of Gore's Island, thence southwesterly between the Aleutian Islands and Asia. Some authorities place it east of the Behring Islands. It then passes southwesterly some degrees east of Cape Lopatka, and the group of the Kurile Islands, thence just east of Japan, keeping west of Guadalupa and Marguerite's Islands, but east of Bonin, Liu Kiu Islands, and south east of Formosa.

"The line then passes through Bashee Channel just north of Bashee Island, and enters the China Sea east of Hong Kong, then passes south just west of the Philippine Islands, but keeping east of Palawan Islands. It is here that it reaches its most western point, being about 116 degrees east longitude. It then takes a south easterly course through the Sulu Sea south of Mindanao Island and north of Gilolo Island. Thence it passes, nearly parallel to the equator and just north of it, to a point about 165 degrees east longitude, just north of Schank's Island. From here it goes south easterly, leaving High Island, Gilbert's Archipelago, Taswell's Island, and the DePeyster group on the north east, thence past the Navigator's or Samoan Islands, to longitude 168 west. It then turns south, keeping east of the Friendly, Tonga, Vasquez, and Curtis Islands, and west of the Society and Cook's Islands; thence it continues, bearing a little to the west, so as to cross, according to some authorities, Chatham's Island, and thence to the south pole."

We have now studied the earth, the moon, and the sun in their simplest relations to each other, and their most important points of interest to us. This work will be robbed of its greatest importance if you do not work out all of the problems given, which can be easily done by the assistance of your Tellurian.

You will find in the "Appendix" some things that will interest you, and some definitions that you should study from time to time as you have need of them.

APPENDIX.

SOME FIGURES.

The earth's diameter is about 7,925½ miles at the equator.

The moon's diameter is 2,160 miles.

The sun's diameter is 852,584 miles.

The moon's nearest point to the earth is 225,719 miles.

The moon's most distant point from the earth is 251,947 miles.

The earth's nearest point to the sun is 89,894,951 miles.

The earth's most distant point from the sun is 92,965,489 miles.

The moon travels in her orbit at the rate of 2,273 miles per hour.

The earth travels in her orbit at the rate of 65,533 miles per hour.

To get a better idea of the magnitude of these

distances, remember that our fastest express trains do not average 60 miles per hour. See how long it would take to cross the sun's disc.

It is 477,666 miles across the moon's orbit. This is 374,918 miles less than the sun's diameter.

SOME DEFINITIONS.

Apparent Rotation of the Heavens.—If on any clear night you watch the heavens for a few hours, you will notice that the stars have changed their positions, not with reference to each other, for they present the same figures for centuries, but they seem to be crossing the heavens from east to west. Those that were near the horizon and east of the point of observation appear higher up in the heavens, while those west of the point of observation have disappeared below the horizon. All of the stars seem to move from east to west, except those near the Pole Star, which describe a circle around it. This is called the *Apparent Rotation of the Heavens*. The rotation of the earth causes all points in the heavens to appear to be moving, while, in fact, they have not greatly changed their relative position for ages.

The following figure will show why the revolution of the earth causes a star to rise, ascend to the zenith, and disappear below the western horizon:

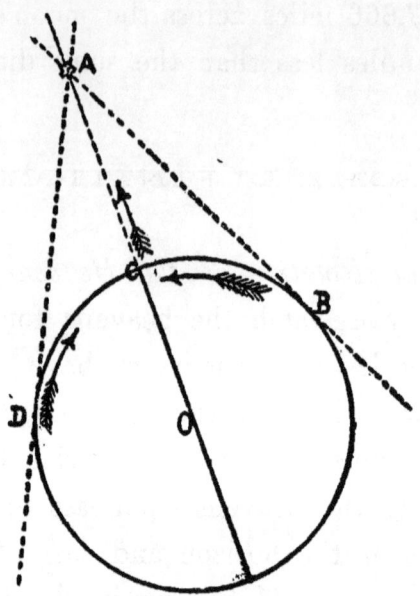

The circle, of which O is the center, represents the earth, and B, C, and D are points at which observers, at the same moment, are watching the star at A. The direction from B to D is east, and from D to B is west, while the star A is directly over the head of the observer at C. The star will appear to be rising to the observer at B, and setting to the observer at D, and, in looking at it, they will look in the direction that the arrows point. As the earth rotates from B to C, the line AB will gradually take the direction of AC; and, as the earth continues her rotation, it will carry the point B to the point D, and then the line AB will take the same direction as the line AD, and the star will appear to have moved from the eastern to the western horizon.

Cardinal Points.—The four principal points of the compass,—*North, East, South,* and *West.*

Circle.—A ring; the compass or circuit of any thing or place; a single curved line, every point of which is equidistant from a point within called its center. *Polar Circles* are the Arctic and Antarctic circles. *Great Circles* are those circles whose planes pass through the center of a sphere and divide it into two equal parts.

Celestial Meridian.—Is the great circle that passes through the zenith and the poles of the heavens. Its plane will also pass through the poles and the center of the earth. I have called it the *Noon Meridian* and the *Sun's Meridian.*

Celestial Sphere.—Is the vast, hollow expanse of the heavens. It appears to us like the inner surface of a great dome, and the stars look like gilded specks attached to it.

Ecliptic.—The apparent path of the sun in the heavens for a year,—so called because an eclipse cannot take place unless the moon be in or near the ecliptic.

Equator.—The supposed or imaginary great circle which passes around the earth at an equal distance from both poles, and divides the earth into the Northern and the Southern Hemispheres.

Equinox.—The time when the sun crosses the equinoctial line, and when all places on the earth's surface have days and nights of equal length. The *Vernal Equinox* occurs on the 20th of March,* at the moment when the sun crosses the equator. The celestial meridian that contains this point is the meridian from which astronomers reckon the location of the stars, the same as we reckon longitude from the meridian of Greenwich. The *Autumnal Equinox* occurs September 22d, when the sun again crosses the equator; but it is not used as a point from which to reckon locations. In reckoning from the Vernal Equinox, astronomers always reckon eastward, and completely around the circle.

Hemisphere.—Half of a sphere. The sun shines on one-half of the earth's surface all of the time, and I have called this half the Day Hemisphere. The

* The dates of the equinoxes and solstices may not be the same for all points on the surface of the earth, because of their difference in time. Suppose that the sun should cross the equator at the meridian of Greenwich at 2:30 o'clock A. M. on the 21st of March. Its time of crossing, for the United States, would be at some time before midnight on the 20th of March. The sun does not cross the equator at the same point, or at the same time each year.

half of the earth's surface that is in the shadow is the Night Hemisphere.

Hour-Circles.—The twenty-four meridians represented on the surface of the globe. They are fifteen degrees apart, or just one hour in time. Strictly speaking, hour-circles are meridians in the celestial sphere that are fifteen degrees apart, and located with reference to the celestial meridians.

Meridians.—Great circles that pass through the poles of the earth or of the heavens. These lines are imaginary, and every point on the earth's surface may be considered to have a meridian passing through it. Only twenty-four of them are represented on the globe, and the meridian that passes through Greenwich is the one from which the longitude of all places on the earth is reckoned.

Parallels.—Circles that cut the meridians at right angles. Only one of them is a great circle, the *Equator;* and, as with the meridians, every point on the surface of the earth may be considered to have a parallel passing through it.

Rotation and Revolution.—Revolution is a broader term than rotation, and includes all that is meant by it, while rotation is more limited in its meaning. *Rotate* means to revolve around a point that is the center of the revolving body; and *Revolve* may mean to rotate, or it may mean to go around a point that is not the center of the revolving body.

Solstices.—Points in the earth's orbit where the sun has reached his most northern or southern latitude. There are two,—the *Summer Solstice*, which indicates the most northern latitude of the sun, and the *Winter Solstice*, which indicates his most southern latitude.

Tropics.—The parallels of latitude in which the solstices occur. They are about $23\frac{1}{2}$ degrees north and south of the equator and parallel to it. They mark the limits in which the sun moves in his yearly course. The one north of the equator is called the *Tropic of Cancer*, and the one south of the equator the *Tropic of Capricorn*.

Zodiac.—An imaginary belt, about sixteen degrees wide (about eight degrees each side of the ecliptic), within which lie the planes of the orbits of all the planets. It is divided into twelve signs. (See Chapter IX.)

A LESSON ON THE ALMANAC.

On the next page is given a cut of the Zodiac, together with the characters that represent the zodiacal signs. The characters that represent the sun, moon, and the planets, are given, with the signs for conjunction, quadrature, and opposition. The moon is said to be in *Apogee* (*ge* means "earth") when she is at the most distant point in her orbit from the earth, and to be in *Perigee* when she is at the nearest point in her orbit to the earth. The earth is said to be in *Aphelion* when she is farthest from the sun, and in *Perihelion* when she is nearest the sun. The signs for the ascending and descending nodes are given. Remember that *node* means the *point* where the moon crosses the sun's path.

STANDARD TIME.
The calculations of this Almanac are given in local time. In places where what is now called standard time has been substituted for local time, our values can be changed to standard time by applying a correction found as follows: For any place *east* of one of the standard meridians, and taking that meridian's time, four minutes is to be *subtracted* for every degree of difference of longitude; and for any place *west* of the meridian four minutes for each degree of difference is to be *added*.

APPARENT RELATIVE POSITION OF THE EARTH, THE SUN, THE MOON, AND THE SIGNS OF THE ZODIAC.

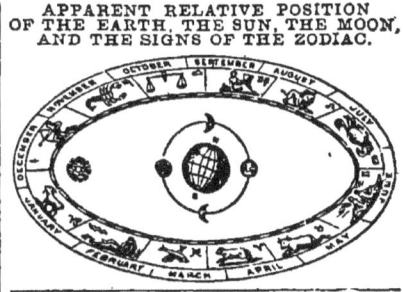

CHRONOLOGICAL CYCLES.
Domin. Letter . . G | Solar Cycle   27
Epact . . . . 23 | Roman Indiction,   7
Golden Number . 14 | Julian Period . . 6607

ZODIACAL. — ♈ ARIES — The Ram, Head and Face; ♉ TAURUS — The Bull, Neck; ♊ GEMINI — The Twins, Arms; ♋ CANCER — The Crab, Breast; ♌ LEO — The Lion, Heart; ♍ VIRGO — The Virgin, Bowels; ♎ LIBRA — The Balance, Reins; ♏ SCORPIO — The Scorpion, Secrets; ♐ SAGITTARIUS — The Bowman, Thighs; ♑ CAPRICORNUS — The Goat, Knees; ♒ AQUARIUS — The Waterman, Legs; ♓ PISCES — The Fishes, Feet.

CHARACTERS. — ☉, The Sun; ☿, Mercury; ♀, Venus; ⊕, The Earth; ☾, The Moon; ♄, farthest north; ♃, farthest south; ♂, Mars; ♃, Jupiter; ♄, Saturn; ♅, Uranus; ♆, Neptune; ☌, Conjunction; ▢, Quadrature, 90° from ☉; ☍, Opposition, 180° from ☉; Apogee, far from ⊕; Perigee, near ⊕; Aphelion, far from ☉; Perihelion, near ☉; ☊, Ascending Node; ☋, Descending Node.

Mercury (☿) will be evening star about February 25, June 23, and October 19; and morning star about April 10, August 8, and November 27.
Venus (♀) will be evening star till February 15; then morning star till November 30; and then evening star again the rest of the year.
Jupiter (♃) will be evening star till June 4; then morning star till December 22; and then evening star again the rest of the year.

APPENDIX.

The following table is to be used in the study of Chapter XIV:

First Month. JANUARY, 1894. 31 Days.

	MOON'S PHASES.				CALENDAR			Sun Slow.	CALENDAR		
	Halifax. D. h. m.	Montreal. D. h. m.	Detroit. D. h. m.	St. Paul. D. h. m.	For Northern N. E. and N. Y.; Upper Michigan, Wis., and Minn.; No. Dakota; Montana and Washington.				For Massachusetts; So. N. H. and Vt.; Cent. N.Y.; Lower Mich., Wis., and Minn.; So. Dak.; Idaho and Oreg'n.		
New M.	6 10 53 A.	6 10 13 A.	6 9 35 A.	6 8 55 A.							
First Q.	14 7 55 A.	14 7 15 A.	14 6 37 A.	14 5 57 A.							
Full M.	21 10 57 M.	21 10 17 M.	21 9 39 M.	21 8 59 M.							
Last Q.	28 0 37 A.	28 11 57 M.	28 11 19 M.	28 10 39 M.							
D. M.	D. W.	Miscellaneous Phenomena.			☾'s pl. Const.	Sun rises. h. m.	Sun sets. h. m.	Moon rises. h. m.	m.	Sun rises. h. m.	Sun sets. h. m.	Moon rises. h. m.
1	Mo	Circumcision. Slave trade abol. 1808.			♓	7 41	4 27	2 42	4	7 30	4 39	2 35
2	Tu	♀☊. Dr. Ure d. 1857. *Snow.*			♓	7 41	4 28	3 50	4	7 30	4 39	3 40
3	We	☌♂☽. Bat. Princeton, 1777.			♈	7 41	4 29	4 56	5	7 30	4 40	4 44
4	Th	Bombardment of Paris, 1871. *Cold.*			♈	7 41	4 30	6 0	5	7 30	4 41	5 46
5	Fri	☌☿☽; ☽ in apog. *Blustering.*			♉	7 41	4 31	6 59	6	7 30	4 42	6 44
6	Sat	Epiphany. Mme. D'Arblay d. 1840.			♉	7 41	4 32	sets	6	7 30	4 43	sets
1) 1st Sunday after Epiphany.					*Venus in Aquarius.*			*8h. 52m. Day's length. 9h. 15m.*				
7	Su	6. The French in Mexico, '62.			♊	7 41	4 33	4 50	7	7 29	4 44	5 3
8	Mo	Eli Whitney d. 1825. *Clear*			♊	7 40	4 34	5 56	7	7 29	4 45	6 7
9	Tu	Conn. adopted constitution, 1788.			♋	7 40	4 35	7 4	8	7 29	4 46	7 12
10	We	☌♀☽; ♀ gr. brilliancy. *and cold.*			♋	7 40	4 36	8 12	8	7 29	4 47	8 18
11	Th	☿ in aphel. F. Scot Key d. 1843.			♌	7 39	4 37	9 21	8	7 29	4 48	9 24
12	Fri	Terrible Dakota blizzard, 1888.			♌	7 39	4 39	10 30	9	7 28	4 49	10 30
13	Sat	☌☽☊. Salm. P. Chase b. 1808.			♍	7 38	4 40	11 40	9	7 28	4 50	11 37
2) 2d Sunday after Epiphany.					*Mars in Scorpio.*			*9h. 4m. Day's length. 9h. 24m.*				
14	Su	14. ☐♄☉. Car. Manning d.'92.			♎	7 38	4 42	morn	9	7 28	4 52	morn
15	Mo	♃ stat. Dr. Parr b. 1747.			♎	7 37	4 43	0 52	10	7 27	4 53	0 47
16	Tu	☌♃☽. Gen. Hazen d. 1887.			♏	7 36	4 44	2 10	10	7 27	4 54	2 2
17	We	☌♅☽. Geo. Bancroft d. 1891.			♏	7 36	4 45	3 32	10	7 26	4 55	3 21
18	Th	German empire proclaimed, 1871.			♐	7 35	4 47	4 54	11	7 25	4 56	4 40
19	Fri	Gen. Zollikoffer killed, 1862.			♑	7 35	4 48	6 9	11	7 25	4 58	5 54
20	Sat	☽ in perig. King Kalakaua d. 1891.			♑	7 34	4 49	7 13	11	7 24	4 59	6 58
3) Septuagesima Sunday.					*Jupiter in Aries.*			*9h. 17m. Day's length. 9h. 37m.*				
21	Su	21. Geo. D. Prentice d. 1870.			♒	7 33	4 50	rises	12	7 23	5 0	rises
22	Mo	Byron b. 1788. *Snow-*			♒	7 32	4 52	6 15	12	7 23	5 1	6 23
23	Tu	♀ stat. Gustave Doré d. 1883.			♓	7 31	4 53	7 33	12	7 22	5 3	7 43
24	We	Rev. Charles Kingsley d. 1875. *storm.*			♓	7 30	4 55	8 57	12	7 21	5 4	8 57
25	Th	Conversion of St. Paul.			♈	7 29	4 56	10 9	13	7 21	5 5	10 8
26	Fri	☌♂☽; ☌♂☽. Dr. Jenner d. 1823.			♈	7 28	4 58	11 20	13	7 20	5 6	11 16
27	Sat	☌♄☽. Emp. William II. b. 1859.			♉	7 27	4 59	morn	13	7 19	5 8	morn
4) Sexagesima Sunday.					*Saturn in Virgo.*			*9h. 35m. Day's length. 9h. 51m.*				
28	Su	28. ☌☉☽. Stanley b. 1841.			♊	7 26	5 1	0 29	13	7 18	5 9	0 23
29	Mo	☌♀☉ sup. *Cold.*			♊	7 25	5 2	1 39	13	7 17	5 10	1 30
30	Tu	♀ gr. hel. lat. S. Osceola d. 1838.			♋	7 24	5 4	2 48	14	7 17	5 12	2 36
31	We	☌♂☽. Rev. C. H. Spurgeon d. 1892.			♋	7 23	5 5	3 53	14	7 15	5 13	3 39

MODEL LESSON ON CHAPTER XIV.

Q. At what point on the earth's surface will the attraction of the sun or of the moon be the strongest?

A. If a line be drawn through the center of the earth and the center of the sun or of the moon, the point will be found in this line and at the surface of the earth.

Q. Are the tides at this point higher than elsewhere on the same side of the earth?

A. If the surface of the earth were an unbroken expanse of water, they would be higher at some distance east of this point, and the movement of the tides would be quite different from what it is now. As it is, the coast lines have a disturbing influence on the tides, not only retarding their movements, but changing their direction and height.

Q. What is the height of the tides in mid-ocean?

A. Perhaps less than two feet. There the height of the tides is the result of the attraction of the sun, or of the moon, or both, undisturbed by other influences.

Q. Why is not the height of tides the same at all places? If the cause is constant, why is not the height uniform?

A. The length and outline of seacoasts are the prime cause of their variation. If the shape of the coast is such as to force the tide into a corner, its height will be greatly increased, as is the case in the Bay of Fundy. While the cause of the tides is constant, it does not act with equal force at all times during the month, the reason for which is evident.

Q. Do tides travel with the same rate of speed all over the surface of the earth?

A. The extent and the depth of the bodies of water where tides are formed modify the rate at which they travel around the earth. Since North and South America extend almost from pole to pole, they act as an effectual barrier to the tides formed in the Atlantic Ocean. The tides beginning with the western coast of the Americas travel northwest across the Pacific at the rate of about 850 miles an hour. The tidal wave in the shallow waters of Oceania travels at the rate of from 400 to 600 miles an hour.

Q. Can the time at which the tides will reach any given point be reckoned?

A. Yes; the tides depend upon conditions that are constant, so far as any location is concerned; but different locations are surrounded by different conditions.

Q. How does it happen that tides occur at opposite points on the earth's surface?

A. This is a different question. By assuming that the theory given on page 112 is correct, it is easy to see that the attraction of the moon would make a tidal wave on the portion of the earth's surface that is toward the moon. In the same way the solid portions of the earth are attracted by the moon, and this attraction would, probably, lessen the force of gravity on the opposite side of the earth from the moon. Now, water is slightly compressible, and, where there is great depth, it exerts an enormous pressure on the lower strata, so that they are more dense than the upper. If this pressure is relieved the whole body will be less dense, and will occupy more room. This would produce a wave unless the force of gravity were uniformly lessened all over the earth's surface, which is contrary to the theory.

Q. Are there other theories for the causes of tides?

A. Yes. Perhaps you will find one in your Physical Geography, or Encyclopædia. Look it up carefully.

Q. What is meant by the "moon's node"?

A. It is the point in the moon's orbit where the moon's path cuts the ecliptic. The point where the moon passes from the southern portion of the ecliptic to the northern portion is called the ascending node. The opposite node is called the descending node.

Q. Can you tell whether the moon is north or south of the ecliptic?

A. Yes; the convex side of the moon's crescent is always toward the ecliptic,—that is, toward the sun. Look at the old moon at 3 o'clock in the morning, and keep the above in mind; you will be surprised.

Q. Why does the moon appear to be so much farther north at some times than she does at others?

A. Because we compare her position with the horizon. She follows about the same path in the heavens, from month to month, during the year. She never gets more than about five* degrees further north or five degrees further south than the sun does. So it must be our relative position that changes from time to time as the observations are made. By the earth's rotation points on the equator are carried around a circle something less than 25,000 miles in circumference. This circle is more or less inclined to the ecliptic, according to the time of the year that the observations are being made; and viewing the sun and the moon from different points has a wonderful effect upon their apparent positions. The experiments with the Tellurian for finding the apparent direction of the sun for different times of the day and of the year, can be made for finding the apparent direction of the moon.

* 5° 8′ 40″.

Q. How can the ecliptic be located in the heavens?

A. By learning the location of the planets. This can be done by studying the moon's movements as directed on page 122. The planets are near the ecliptic, and a line passing through them will indicate nearly the sun's and the moon's path in the heavens. You can see that the position of the moon at midnight on a midwinter day would be almost the same as that of the sun at noon on midsummer's day.

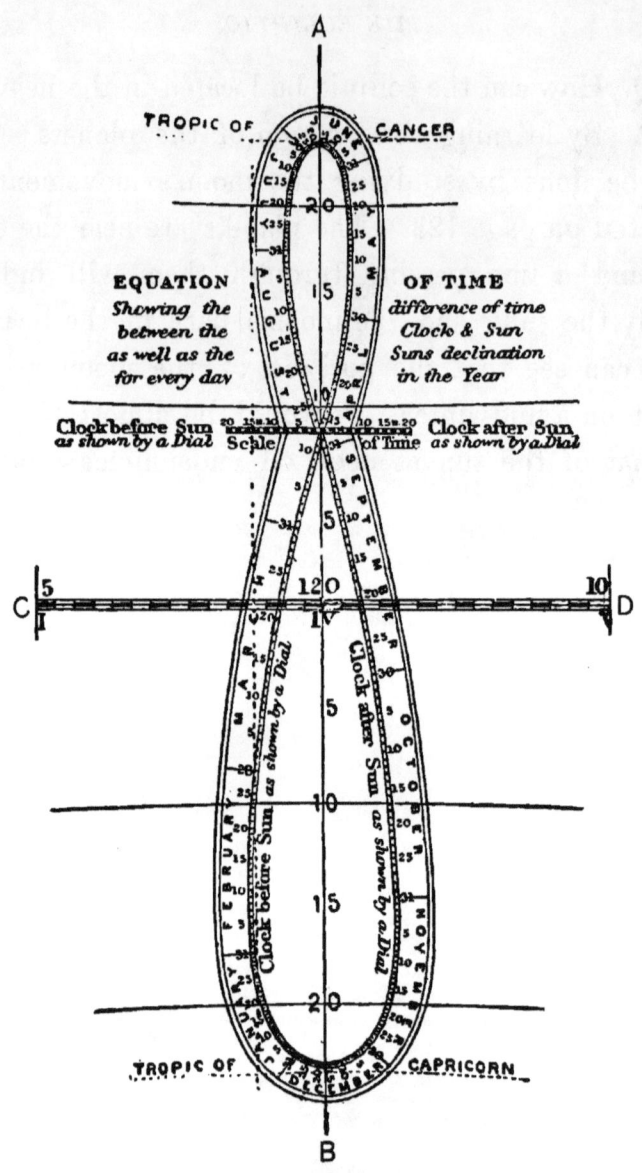

THE ANALEMMA.

THE ANALEMMA.

The foregoing figure is a representation of the *Analemma* as shown on the Tellurian. AB is a meridian, and the figures along this meridian indicate degrees of latitude. CD is a section of the equator. Various other parts of the figure are named in the cut.

Q. For what is the analemma used?

A. The analemma is used to show the latitude at which a line drawn through the centers of the sun and the earth would fall upon the earth's surface for any day of the year. This line would represent the direct rays of the sun.

Q. Can we truly say that this is the latitude of the sun?

A. It amounts to the same thing, but the sun is not spoken of as having latitude. The term "declination" is used in its stead, meaning distance in degrees north or south of the celestial equator.

Q. Are, then, the latitude of the direct rays of the sun and the sun's declination the same?

A. Yes.

Q. The declination of the sun on the 26th of

February is about the same as on the 15th of October. Why is it not so warm, in northern latitudes, on the former date as on the latter?

A. Each acre of ground in the same latitude and having the same slope receives the same amount of heat, but it does not produce the same effect; for, before October 15th, the earth has been receiving the heat of a summer's sun, while prior to February 26th, the earth has been storing up cold in ice-locked rivers and lakes, and snowclad hills and valleys. Much heat must be absorbed in overcoming this vast amount of cold; so the winter's cold is not all chased away until the sun has crossed the equator, which does not occur until the latter part of March.

Q. What else can we learn from the analemma?

A. We can more readily see the distance in latitude passed over by the direct rays of the sun in any month of the year, and compare this with the distance passed over in any other month. (Compare June and September.)

Q. Why is this distance greater for some months than it is for others?

A. Because the rate of the earth's movement in her orbit is not uniform, the form of her orbit being that of an ellipse; and, in the case of June and December, the sun seems to turn in his course north or south and go in the opposite direction, so that the distance in latitude passed through in these months is much less than it is for any of the other months. A close observation of the analemma at these points will make more clear the meaning of the word "Solstice." (See page 63.)

Q. What is meant by the "equation of time"?

A. One of the principal means that we first had of reckoning time was by observing the sun, calling it noon when the sun was on our meridian, the highest point in the heavens for the day. Then came the division from noon to noon into hours and minutes, and the attempt to get some mechanical device that would record them as they occurred. This brought into existence the clock. It was found that the sun did not reach the same meridian at exactly the same time each day, and that the time as indicated by the sun-dial and mean solar time and that kept by good timepieces did not agree, except at four different dates during the year. The greatest difference occurs on the 2d of November, and about the 11th of February. On the 2d of November the sun is 16 minutes and 20 seconds fast, and, on the 10th and 11th of February, 14 minutes and 27 seconds slow.

This difference is called the equation of time, and in some almanacs it is given for each day of the year.

172   APPENDIX.

MOON'S PHASES.			BOSTON.			PITTSBURGH.			NEW ORLEANS.		
			D. H. M.			D. H. M.			D. H. M.		
● New Moon,			5 5 1 Eve.			5 4 25 Eve.			5 3 45 Eve.		
☽ First Quarter,			13 5 59 Mor.			13 5 23 Mor.			13 4 43 Mor.		
○ Full Moon			19 9 32 Eve.			19 8 56 Eve.			19 8 16 Eve.		
☾ Last Quarter,			27 7 44 Mor.			27 7 8 Mor.			27 6 28 Mor.		

Month.	Week.	HISTORICAL EVENTS.	Moon's Cot.	Sun Slow. M. S.	Sun rises. H. M.	Sun sets. H. M.	Moon rises. H. M.	Sun rises. H. M.	Sun sets. H. M.	Moon rises. H. M.	Sun rises. H. M.	Sun sets. H. M.	Moon rises. H. M.
1	Th	Sir Edward Coke b..1552	♏	13 51	7 14	5 14	4 38	7 10	5 18	4 31	6 51	5 38	3 54
2	Fri	Rich.-H. Dana died,1879	♐	13 59	7 13	5 16	5 31	7 9	5 19	5 24	6 50	5 38	4 46
3	Sa	F. W. Robertson b., 1816	♐	14 5	7 12	5 17	6 17	7 8	5 20	6 10	6 49	5 39	5 35

(5.) QUINQUAGESIMA—SHROVE SUNDAY.   Luke 18.   Day's Length, (Pitts.) 10 h. 14 m.

4	S	Battle Moorefield, 1864	♑	14 11	7 11	5 18	6.54	7 7	5 21	6 48	6 49	5 40	6 17
5	Mo	Mass. rat. Constitu., 1788	♒	14 15	7 10	5 19	sets	7 6	5 22	sets	6 48	5 41	sets
6	Tu	Ft. Henry captured,1862	♒	14 19	7 9	5 21	6 10	7 5	5 24	6.15	6 47	5 42	6 34
7	We	Pitt's Cabinet diss.. 1801	♒	14 22	7 7	5 22	7 16	7 4	5 25	7 19	6 47	5 42	7 31
8	Th	New Prussian const.1847	♓	14 25	7 6	5 23	8 23	7 3	5 26	8 25	6 46	5 43	8 29
9	Fri	Gen.·Hancock died, 1886	♓	14 26	7 5	5 25	9 30	7 2	5 27	9 30	6 45	5 44	9 27
10	Sa	Reverdy Johnson d. 1876	♓	14 27	7 4	5 26	10 39	7 0	5 29	10 38	6 44	5 45	10 27

(6.) 1st SUNDAY IN LENT.   Matt. 4.   Day's Length, (Pitts.) 10 h. 31 m.

11	S	Daniel Boone born, 1735	♈	14 27	7 2	5 27	N 50	6 59	5 30	11 47	6 44	5 46	11 29
12	Mo	O'Brien & Dillon sur.,'91	♈	14 26	7 1	5 29	MOR.	6 57	5 32	MOR.	6 43	5 47	MOR.
13	Tu	Wm. and Mary proc.1689	♈	14 24	6 59	5 30	1 5	6 56	5 33	1 0	6 42	5 47	0 35
14	We	Gen. Sherman died, 1891	♉	14 22	6 58	5 31	2 22	6 55	5 34	2 16	6 41	5 48	1 43
15	Th	Louis XV. born, 1710	♉	14 19	6 57	5 32	3 36	6 54	5 35	3 29	6 40	5 49	2 52
16	Fri	St. Independence lost'53	♊	14 15	6 56	5 34	4 43	6 52	5 36	4 36	6 39	5 50	3 58
17	Sa	Thiers elected Pres.,1871	♊	14 10	6 54	5 35	5 39	6 51	5 37	5 33	6 38	5 51	4 58

(7.) 2d SUNDAY IN LENT.   Matt. 15.   Day's Length, (Pitts.) 10 h. 48 m.

18	S	Charleston captured, '65	♋	14 5	6 53	5 36	6 22	6 50	5 38	6 17	6 37	5 51	5 49
19	Mo	Aaron Burr arrest., 1807	♋	13 59	6 51	5 37	RISE.	6 49	5 39	RISE.	6 36	5 52	RISE.
20	Tu	David Garrick born,1716	♌	13 52	6 50	5 39	6 30	6 47	5 40	6 33	6 36	5 53	6 45
21	We	Wash. Mon. dedicat.1885	♌	13 45	6 48	5 40	7 43	6 46	5 42	7 44	6 35	5 54	7 48
22	Th	David II., Scot'l'd, d..1721	♍	13 37	6 47	5 41	8 55	6 44	5 43	8 55	6 34	5 54	8 50
23	Fri	Hornet capt. Penguin'15	♍	13 29	6 45	5 43	10 5	6 43	5 44	10 3	6 33	5 55	9 51
24	Sa	Revolution in Mex., 1821	♎	13 19	6 44	5 44	11 14	6 41	5 45	11 11	6 32	5 56	10 51

(8.) 3d SUNDAY IN LENT.   Luke 11.   Day's Length, (Pitts.) 11 h. 6 m.

25	S	1st U. S. Bank chart..1791	♎	13 10	6 42	5 45	MOR.	6 40	5 46	MOR.	6 31	5 56	11 50
26	Mo	French Republic, 1848	♎	13 0	6 41	5 46	0 22	6 38	5 48	0 17	6 30	5 57	MOR.
27	Tu	Tories defeat'd N C1778	♏	12 49	6 39	5 47	1 27	6 37	5 49	1 21	6 29	5 58	0 48
28	We	Rachel born, 1820	♏	12 37	6 38	5 49	2 28	6 35	5 50	2 21	6 28	5 58	1 45

Above is a table for the month of February, 1894. The column in which the equation of time is given is marked "Sun Slow."

Q. Can this equation of time be determined by the Tellurian?

A. Yes; within a few seconds. Fix the movable meridian on your Tellurian so that it will be in the position of the dotted line in the cut of the analemma, on page 168, and notice that it touches February at about the 11th. Also notice that it crosses the scale of time between 14 and 15, which shows that the difference in time is between 14 and 15 minutes, and that it is on the side of the analemma that indicates that the sun is slow compared with the clock.

Notice the column in your almanac that reads "Sun Fast" or "Sun Slow."

FINIS.

POLITICO-RELIEF MAPS,

showing the contour and topography of a country, as well as the political divisions and other principal details.

All teachers and students recognize the importance of studying the topography of the land as well as the political boundaries, location of physical features, such as lakes, rivers, etc.

It has been difficult heretofore to pursue such studies without necessarily referring to one map for the political features, to another for the physical features, and to a third for the topography, or, possibly, to the imagination, or a crude and imperfect relief map made by the teacher at odd moments. These Maps happily overcome all former difficulties, because they combine the relief features with all the features of the ordinary flat map, the names, etc., appear in print, the same as on any flat map. The most wonderful feature of these Maps is—How is it possible to have the printing so regular, so perfect, when the surface is so rough and uneven, representing, as it does, the appearance of the country? The substance is light; not including the frame that surrounds it, the weight will not exceed five or six ounces, and yet it is impossible to batter down the surface by striking it with a hammer.

Being in a position to feel the pulse of the leading educators generally upon topics that pertain to their profession, and having realized that the time *has* come when relief maps must take the place of all flat maps, the publishers take great pleasure in announcing to their many friends and patrons that, after years of experimental work, and an outlay of many thousands of dollars, they have succeeded in producing the **First Complete Set of Politico-Relief Maps;** gaining for themselves the highest praise, for their sagacity and enterprise, from the leading educators of the world. These Maps have been pronounced one of the marvels of the Nineteenth Century, by the most prominent Geographers and Scientists, and interest the teacher, the pupil and parent, and in every instance call forth expressions of surprise and astonishment from all.

The complete set of seven Maps—consisting of North America, South America, Europe, Asia, Africa, United States, and The World (Mercator's Projection)—are each mounted in a beautiful carved Solid-Oak Frame, and enclosed in an elegant Antique-Oak Cabinet Case.

Full particulars given, and illustrated circulars sent to any address.

Central School Supply House,

S. E. Cor. Monroe and Fifth Ave., = Chicago, Ill.

The Teachers' Anatomical Aid.

It is a series of lithographic plates, 25x42 inches in size, and complete manikins of the human body—the head, the eye, the ear, the tooth, the lungs and heart, and the larynx; illustrating human anatomy and physiology, also showing the effects of alcoholic drinks and narcotics on the system.

The Aid has been before the public for a number of years and has merited the highest praise, and stands pre-eminently superior in its line. The large sale it has enjoyed, and its use in the best schools bespeak for it the careful consideration of all who may be interested in apparatus of this kind. The following is a list of the large sheets, with full-sized illustrations, and what they teach:

1. Title Page.
2. Front View of Skeleton.
3. Back View of Skeleton.
4. The Muscular System.
5. The Arteries and Veins.
6. The Nervous System; The Five Senses; The Sympathetic System.
7. The Formation and Circulation of Blood; The Oxygenation of Blood; Microscopic Details of Structure.
8. The Stomach in Health. The Primary and Secondary Effects of Alcohol on the Mucous Membrane.
9. The Stomach in Advanced State of Alcoholism; State in Delirium Tremens. Effect of Beer or Gin on the Kidneys. Gin or Hob-Nailed Liver.
10. Alcoholic Effect on the Brain, the Heart, the Nerves, the Eyes, the Arteries and Veins, the Liver and Skin. The Baneful Effects of Cigarette Smoking and its Destruction of Respiratory Organs; also Ulcerous and Cancerous Growth in the Larynx.

By the aid of the manikins, the student will go through a process similar to dissection, and have the whole organism before him, with each in its proper place. Without seeing the apparatus, it would be difficult to comprehend how the interior of the body is exhibited, and how the organs may be taken out one after another, until all have been removed.

The Case in which the apparatus is enclosed stands upright 44 inches, and is supported by an easel at an angle of 75 degrees. It is constructed of well-seasoned ash, beautifully carved, making it both durable and artistic. When not in use, it can be closed up, thus protecting it from dust or other injury. The case holds the sheets flat and neat, and does not permit them to twist or roll.

Correspondence solicited. Illustrated circulars and full particulars given on receipt of request. Address

CENTRAL SCHOOL SUPPLY HOUSE,
S. E. Cor. Monroe and Fifth Ave., Chicago, Ill.

The Progressive Reading and Number Study,

BY

MISS MARY E. BURT

is placed before the educational public by the publishers, in the confidenᴜ ɢᴇɴᴇɪ that teachers in every grade of school work will hail its appearance with happy enthusiasm. Believing with all our leading educators, that reading ranks first in importance among the educational studies, they have spared neither care nor money to present, in this Study, a work in keeping with the great importance accorded this subject in the course of instruction given in our schools.

THE ILLUSTRATIONS were selected from some of the most noted masters of the Old and New Worlds. Pen sketches were made from photographs of the originals, with slight modifications, to adapt the picture to the child's capacity, or to emphasize some particular feature.

ART AND ILLUSTRATIONS—The art work alone cost *thousands of dollars*, and the publishers are receiving thanks and congratulations from teachers wherever the Study is shown for the progress they have made.

SUPPLEMENTARY READING is taken up in three books, 6 by 10 inches; they are elegantly illustrated, the pictures affording ample opportunity to enlarge upon the lesson.

ELEMENTARY NUMBER WORK—An urgent demand on the part of our primary teachers, and in all courses of study for our common schools, induced the publishers to combine this course with the earlier lessons in reading. In doing so the inductive method of Grube, with such modifications as have been made in recent years by expert teachers, was followed. This complete course comprises forty pages, 10 by 14 inches in size, and provides both class drill and seat work. There are also several pages devoted to United States money and the operations of making change, etc.

THE COLOR FOLIO—To meet the demand for color work in our schools the publishers have prepared a four-page folio, devoted entirely to the subject of color. The colors used are Prang's; the finest colors in scaled tones yet produced.

TIME is illustrated by means of five lithographed figures, showing: (1) Natural Divisions of Time; (2) Causes Producing Such Divisions; (3) Comparative Time Throughout the World; (4) The Phases of the Moon; (5) Clock Dial with Movable Hands.

THE AUTHOR—The publishers were fortunate in securing the services of Miss Mary E. Burt to prepare and arrange the matter presented in its pages. She has made primary instruction the chief labor and study of her life, and is recognized by educators as one of the best teachers of children in the country.

THE CASE AND MOUNTING—The Main Study contains fifty large pages, 22x35 inches, mounted on heavy linen lined paper, including one page of geometrical forms. It is enclosed in an Antique Ash Cabinet Case, 36 inches high, and arranged so that when twenty-five lessons have been learned, the Main Study can be lifted out and reversed, when the succeeding lessons can be presented as before. The Case stands at an angle of about 75 degrees. When not in use, can be closed and locked, thus protecting it in every way.

Write for illustrated circulars and full particulars to

Central School Supply House,

S. E. Cor. Monroe and Fifth Ave., Chicago.

www.ingramcontent.com/pod-product-compliance
Lightning Source LLC
Chambersburg PA
CBHW031447160426
43195CB00010BB/884